Cutting Edge Techniques in Biophysics, Biochemistry and Cell Biology: From Principle to Applications

Edited by
Anupam Jyoti
Amity Institute of Biotechnology Amity University Rajasthan Jaipur, Rajasthan India

&

Neetu Mishra
Symbiosis School of Biological Sciences
(Formerly called Symbiosis School of Biomedical Sciences)
Symbiosis International (Deemed University), Lavale, Pune, Maharashtra India

Cutting Edge Techniques In Biophysics, Biochemistry And Cell Biology: From Principle To Applications

Editors: Anupam Jyoti and Neetu Mishra

ISBN (Online): 978-981-14-2286-7

ISBN (Print): 978-981-14-2285-0

need for a court order if at any point you breach any terms of this License Agreement. In no event will any delay or failure by Bentham Science Publishers in enforcing your compliance with this License Agreement constitute a waiver of any of its rights.

3. You acknowledge that you have read this License Agreement, and agree to be bound by its terms and conditions. To the extent that any other terms and conditions presented on any website of Bentham Science Publishers conflict with, or are inconsistent with, the terms and conditions set out in this License Agreement, you acknowledge that the terms and conditions set out in this License Agreement shall prevail.

Bentham Science Publishers Pte. Ltd.
80 Robinson Road #02-00
Singapore 068898
Singapore
Email: subscriptions@benthamscience.net

BENTHAM
SCIENCE

CONTENTS

FOREWORD

Biomedical research is gaining prime attention as it is directly affecting human health. Biophysical, biochemical and cellular techniques are the backbone of biomedical sciences; a pre-requisite towards the understanding and treatment of human diseases. In the recent years, biomedical research has provided solutions for several problems faced by human beings. It is an integrated approach that includes various disciplines *viz.* biochemistry, microbiology, genetics and biochemical engineering. After some time, it has been integrated rapidly with new branches of science like molecular and cellular biology, genomics, proteomics, bioinformatics and nanotechnology. Biomedical research involves fundamental scientific principles applied to preclinical understanding of problems to clinical solutions. This book titled 'Cutting Edge Techniques in Biophysics, Biochemistry and Cell Biology: From Principle to Applications', presents a broad overview regarding the basic applications of sophisticated and analytical techniques used in biophysics, biochemistry and cell biology. This book has covers the following areas: *in vitro* cell culture assay, real-time PCR, flow cytometry, and X-ray crystallography, discussed by authors who have quality publications in their proposed chapter area. Additionally each chapter includes application of techniques in specific areas like cells sorting by FACS, disease diagnosis by Real-time PCR, disease modeling under *in vitro* culture and many more. This will help students understand the importance of techniques in biophysics, biochemistry and cell biology research which will set a benchmark for further research.

As a biotechnology scientist, I am happy to recommend this book to the students of universities, both as a text and reference book. The section on technique will be used as textbook and the application section as reference. This book has been written in a way so that it is student-friendly with clean diagrams and protocols of specific techniques. I sincerely believe that the book has been prepared with the scientific skills and will serve as a useful document for the graduate and undergraduate students.

Rajiv Kumar
ICAR-Central Sheep and Wool Research Institute
Avikanagar, Tonk, Rajasthan
India

PREFACE

This ebook titled 'Cutting Edge Techniques in Biophysics, Biochemistry and Cell Biology: From Principle to Applications' provides principles of basic and analytical techniques in biophysics, biochemistry and cell biology and their potential applications in biomedical research. The volume covers the basics of *in-vitro* cell culture techniques, such as flow cytometry including FACS, Real time PCR based disease diagnosis as well as gene expression analysis, X ray crystallography, RNA sequencing and their various biomedical applications, including drug screening, disease model, functional assays, disease diagnosis, gene expression analysis, and protein structure determination. This is a valuable resource for biomedical students, cellular and molecular biologists who want to be an eminent biomedical researcher.

We strongly believe that this book is a reader's delight providing a comprehensive understanding on cell biology, macromolecular structure, *in silico* studies and functioning. It is an excellent introduction for students towards cell biology and biophysics that clearly help them to develop a wider perspective of the field. The idea to encompass knowledge covering from what is known to what is unknown in the writings from the experts in the field results in a concerted effort to justify the various topics. We sincerely hope our efforts will be embraced by students with appreciation and enthusiasm for learning.

Dr. Anupam Jyoti
Amity Institute of Biotechnology
Amity University Rajasthan
Jaipur, Rajasthan
India

&

Dr. Neetu Mishra
Symbiosis School of Biological Sciences (Formerly called Symbiosis School of Biomedical Sciences)
Symbiosis International (Deemed University)
Lavale, Pune, Maharashtra
India

List of Contributors

Anupam Jyoti   Amity Institute of Biotechnology, Amity University Rajasthan, Jaipur, India

Gauri Misra   Amity Institute of Biotechnology, Amity University Uttar Pradesh, Noida, India

Juhi Saxena   Dr. B. Lal Institute of Biotechnology, 6-E, Malviya Industrial Area, Jaipur 302017, India

Nagendra Singh Tomar   School of Biotechnology, Gautam Buddha University, Greater Noida, U.P.-201308, India

Neetu Mishra   Symbiosis School of Biological Sciences (Formerly called Symbiosis School of Biomedical Sciences), Symbiosis International (Deemed University), Lavale, Pune, Maharashtra, India

Roberta Menafra   Laboratoire de Biochimie, École supérieure de physique et de chimie industrielles de la ville de Paris [ESPCI Paris], France

Rohit Saluja   Department of Biochemistry, Medical College Building, 3rd Floor, Saket Nagar, AIIMS Bhopal-462024, India

Sandeep Ameta   ESCPI Paris, 10 rue Vauquelin, 75005 Paris, France

Sanket Kaushik   Amity Institute of Biotechnology, Amity University Rajasthan, Jaipur, India

Swapnil Sinha   Assam Downtown University, Panikhaiti, Guwahati-781026, Assam, India

Vijay Srivastava   Amity Institute of Biotechnology, Amity University Rajasthan, Jaipur, India

Vinod Singh Gour   Amity Institute of Biotechnology, Amity University Rajasthan, Jaipur, India

CHAPTER 1

Animal Cell Culture: From Fundamental Techniques to Biomedical Applications

Neetu Mishra* and **Joyita Banerjee**

Symbiosis School of Biological Sciences (Formerly called Symbiosis School of Biomedical Sciences), Symbiosis International (Deemed University), Lavale, Pune, Maharashtra, India

Abstract: Animal cell culture techniques in today's scenario have become an indispensable tool in the field of biomedical research. It provides a basis to study molecular and biochemical changes associated with disease pathogenesis. It explicitly provides a scope to study gene expressions, regulation, proliferation, and differentiation in normal as well as pathologic conditions. The culturing of animal cells requires aseptic conditions and vital technical skills to carry out successful cell culture experiments. It provides an appropriate model for studying cell and molecular biology, biochemical changes in cells, drug screening and efficacy *etc.* This chapter describes the essential techniques of animal cell culture as well as its applications.

Keywords: Animal cell culture, Adherent cells, Bacterial contamination, Biomedical research, Biomedical applications, Cell counting, Cell lines, Cell viability, Contaminants, Cross-contamination, Cryopreservation, Laboratory instruments, Mycoplasma contamination, MTT assay, Primary cell cultures, Secondary cell cultures, Suspension cells, Thawing of cells, Virus contamination, Yeast contamination.

1. INTRODUCTION

Animal cell culture has become an important tool in biomedical research and applications. Animal cells are the most ideal resources for diagnostics, therapeutics and pharmaceutical applications. Animal cell or tissue culture technique is a rapidly growing field which extends a better understanding of the complex cellular and physiological processes outside the human body in a controlled environment. The term 'animal tissue culture' is used for the technique involved in removing the cells, tissues, or organs from an animal and subsequ-

---

* **Corresponding author Neetu Mishra:** Associate Professor, Symbiosis School of Biological Sciences (Formerly called Symbiosis School of Biomedical Sciences), Symbiosis International (Deemed University), Lavale, Pune, Maharashtra, India; E-mail: nitumishra2007@gmail.com

Anupam Jyoti & Neetu Mishra (Eds.)

ently growing them in an artificial and controlled growing environment consisting of a suitable culture vessel and liquid or semisolid medium supplemented with nutrients and growth factors.

The animal cell culture was first undertaken in 1907 by Ross Harrison, which later on underwent several developments such as the development of antibiotics to avoid contaminations, use of trypsin to remove cells from culture vessels, development of chemically defined cell culture media, laminar flow hoods and development of continuously growing cell lines [1]. These fundamental developments in animal cell culture made it an indispensable tool in cellular and molecular biology. Animal cell culture provides an appropriate model for studying biochemistry and physiology of cells, effects of drugs on the cells, toxicity of compounds, screening of lead compounds in drug development, mutation, and carcinogenesis. Animal cell culture allows obtaining consistent and reproducible results. This chapter, therefore, is aimed to serve as a basic introduction to animal cell culture and its applications in biomedical research.

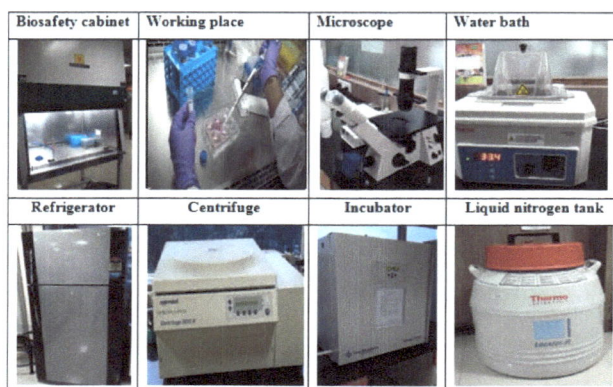

Fig. (1). Basic instruments required in animal cell culture laboratory.

2. FUNDAMENTAL REQUIREMENTS IN ANIMAL CELL CULTURE LABORATORY

The foremost requirement in an animal cell culture laboratory is to adhere to the proper laboratory protocol in order to reduce the chances of contamination in the cell cultures. It is appropriate for the laboratory personnel to know the key concepts about the requisites of tissue culture laboratory and to strictly follow the fundamental guidelines necessary while working in the animal tissue culture laboratory. Table 1 and Fig. (1) summarize the basic materials or reagents and instruments required in the tissue culture laboratory. Fig. (2) depicts the fundamental steps necessary to follow in cell culture laboratory.

Table 1. Essential requisites in animal cell culture laboratory.

A) Materials/ Reagents	B) Instruments
i. Cell culture media with/without phenol red indicator (ready-made/ powdered form) ii. Growth supplements such as fetal bovine serum (FBS), fetal calf serum (FCS) iii. Sterile phosphate buffered saline iv. Essential amino acids such as glutamine v. Cell dissociation reagents such as Trypsin-EDTA (1X), Trypsin phosphate versene glucose (TPVG) (1X) vi. Cell culture flasks or cell culture plates or petridishes vii. Disposable sterile 15ml and 50ml tubes viii. Sterile reagent bottles ix. Disposable sterile serological pipettes (1ml, 5ml, 10ml) x. Pipette controller xi. Gloves xii. 70% ethanol xiii. Tissue rolls	i. 37°C incubator humified with 5% CO_2 ii. Biosafety cabinets iii. Inverted microscope iv. Automated cell counter (alternatively hemocytometer) v. Bench top centrifuge machine vi. Water bath vii. Weighing balance viii. pH meter ix. Refrigerator x. Liquid nitrogen tank

3. TYPES OF CELLS

The cultured cells are usually sub-divided based on their morphologies (shapes or appearances) and functional characteristics [1]. Based on morphologies, cells are classified into three types:

  i. Epithelial-like cells: attached to a substrate, appear flattened or polygonal in shape.
 ii. Lymphoblast-like cells: normally remain unattached to a substrate and remain in suspension with a spherical shape.
iii. Fibroblast-like cells: attached to a substrate and appear elongated or bipolar in shapes, form swirls in heavy cultures.

The functional characteristics of cells depend on their tissues of origin, such as liver, heart, pancreas *etc.* The cultured cells may lose their parental characteristics on being placed in an artificial environment. The biochemical or morphological markers are present to determine whether the cells exhibit similar specialized functions as that of tissues of origin. Some cell lines stop dividing after certain number of divisions. These cell lines are called finite cell lines. Some cell lines divide indefinitely and become immortal, called continuous cell lines. When a finite cell line undergoes a fundamental irreversible change, either intentionally or spontaneously due to drugs, viruses, or radiations, the cells become transformed [1].

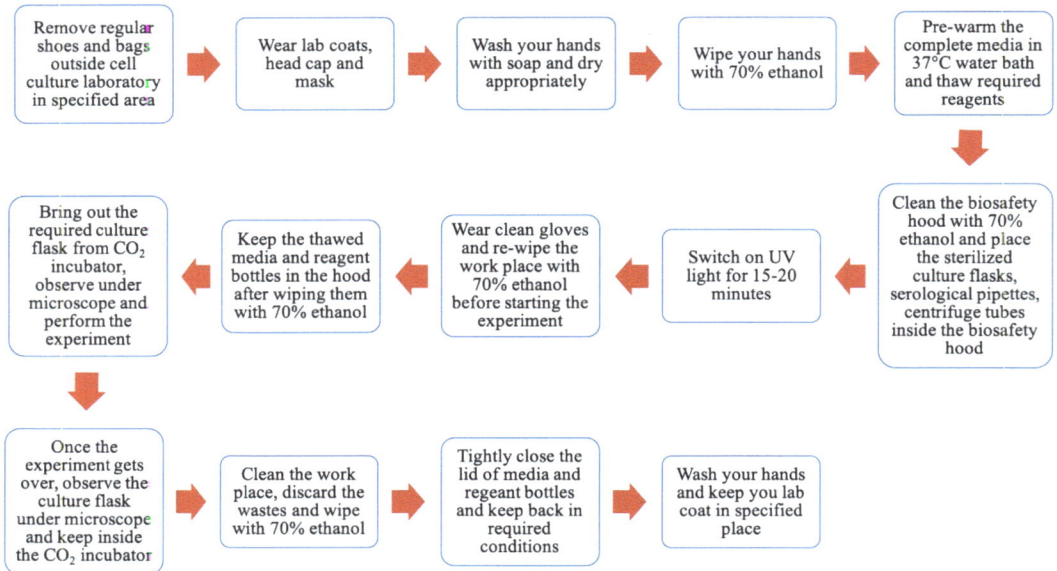

Fig. (2). Fundamental steps to follow in cell culture laboratory.

4. TYPES OF CELL CULTURES

Animal cell cultures are typically of two types: primary and secondary cell cultures. The cells may be directly removed from the tissues or they may be derived from an established cell line or cell strain.

4.1. Primary Cell Culture

When the cells are directly obtained from the cells of a host tissue or organism and placed into a suitable culture environment, where they grow and divide is called primary cell culture. The cells detached from parental tissue grow either as an adherent monolayer or in a suspension, depending on the nature of parent tissues. The primary cell cultures may be done by two methods, either by explants culture or by enzymatic dissociation method. In explants culture method, the first small pieces of tissues are attached to a glass or plastic culture vessel bathed in the culture medium. Subsequently after a few days, individual cells will migrate from the tissue explants into the culture vessel surface where they will start dividing and growing. Next, in enzymatic dissociation method, the cells are disaggregated by using proteolytic enzymes such as trypsin, collagenase. This method is a faster one and more widely used. In this method, a suspension of single cells is created which is then placed into the culture vessels for growing and dividing [1, 2].

4.2. Secondary Cell Culture

When a primary culture is sub-cultured, it is known as secondary culture or cell line or sub-clone. The process involves removing the growth media and dissociating the adhered cells. Sub-culturing of primary cells to different divisions leads to the generation of cell lines. During the passage, cells with the highest growth capacity predominate, resulting in a degree of genotypic and phenotypic uniformity in the population. However, as they are sub-cultured serially, they become different from the original cell [2].

4.3. Maintenance of Cell Cultures or Sub-culturing of Cells

The maintenance of cell cultures in optimal conditions is one of the foremost requisites after obtaining them from cell repositories. Sub-culturing of cells (also known as passaging of cells) enables cells to perform further propagation. Sub-culturing requires removal of the old media from the culture flask and transferring the cells to another culture flask with fresh growth media. The requirement of sub-culturing of cells depends on the confluence of cells. The growth and culturing conditions vary depending on the cell types [2].

4.3.1. Adherent Cell Cultures

Adherent cultures should be sub-cultured when they are in the log phase and have not reached confluence. When the normal cells reach confluence, they stop growing due to contact inhibition. Fig. (**3**) shows the steps required in the sub-culturing of adherent cells.

4.3.2. Suspension Cell Cultures

Sub-culturing of suspension cell cultures also needs to be done once the cells are in log phase and so, they do not reach full confluence. The cells clump together when they reach confluence and the media becomes turbid. The sub-culturing of suspension cell cultures is less complicated as compared to the adherent cultures. Fig. (**4**) shows the steps required in the sub-culturing of suspension cell cultures.

```
┌──────────────┐      ┌──────────────┐      ┌──────────────┐      ┌──────────────┐
│ Remove the   │      │ Wash the     │      │              │      │ Pipette      │
│ old media    │ ──▶  │ cells with   │ ──▶  │ Decant PBS   │ ──▶  │ Trypsin-     │
│ from culture │      │ 2ml PBS      │      │              │      │ EDTA/ TPVG   │
│ flask        │      │ (without     │      │              │      │ (usually 1-2 │
│              │      │ Ca²⁺, Mg²⁺)  │      │              │      │ ml), rotate  │
│              │      │              │      │              │      │ the culture  │
│              │      │              │      │              │      │ flask,       │
│              │      │              │      │              │      │ remove the   │
│              │      │              │      │              │      │ excess       │
└──────────────┘      └──────────────┘      └──────────────┘      └──────────────┘
```

Remove the old media from culture flask → Wash the cells with 2ml PBS (without Ca^{2+}, Mg^{2+}) → Decant PBS → Pipette Trypsin-EDTA/ TPVG (usually 1-2 ml), rotate the culture flask, remove the excess

Collect the cell suspension in a 15 ml centrifuge tube ← Add complete media (usually 2-3 ml) in the culture flask to neutralize trypsin ← Tap the culture flask to dislodge the cells, examine under microscope ← Incubate the culture flask at 37°C for 1-2 minutes

Centrifuge at 200g for 5 minutes → Discard the supernatant → Resuspend the cell pellet in 2-3 ml of fresh media (optional: perform cell counting) → Add the resuspended cell suspension in culture flask containing fresh media

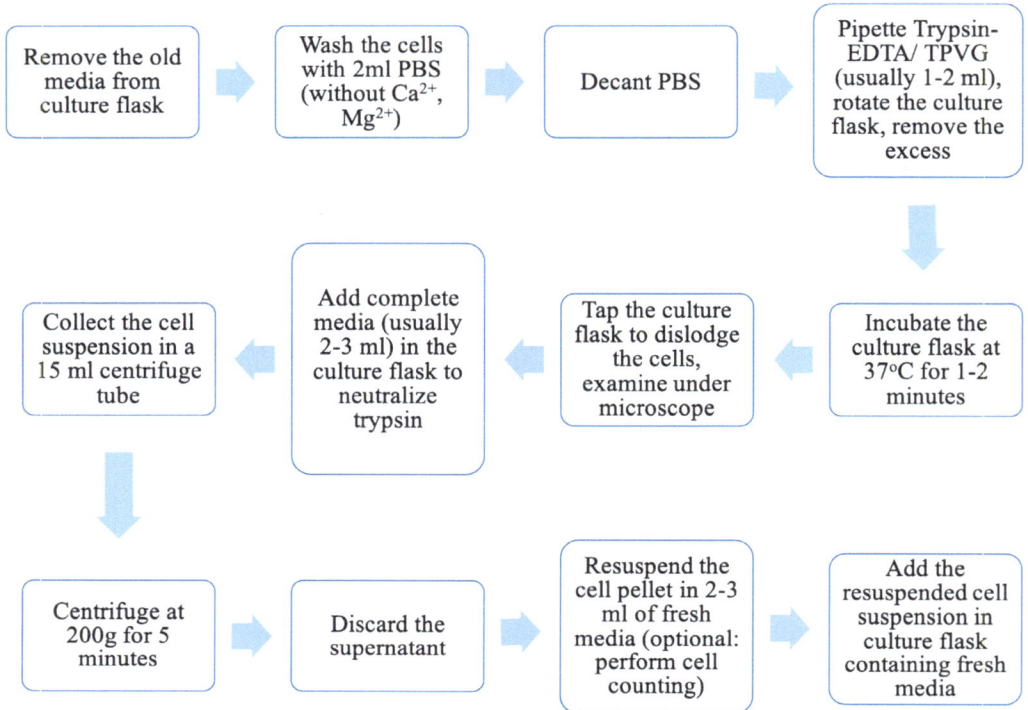

Fig. (3). Flow-diagram showing steps to follow in sub-culturing adherent cell cultures.

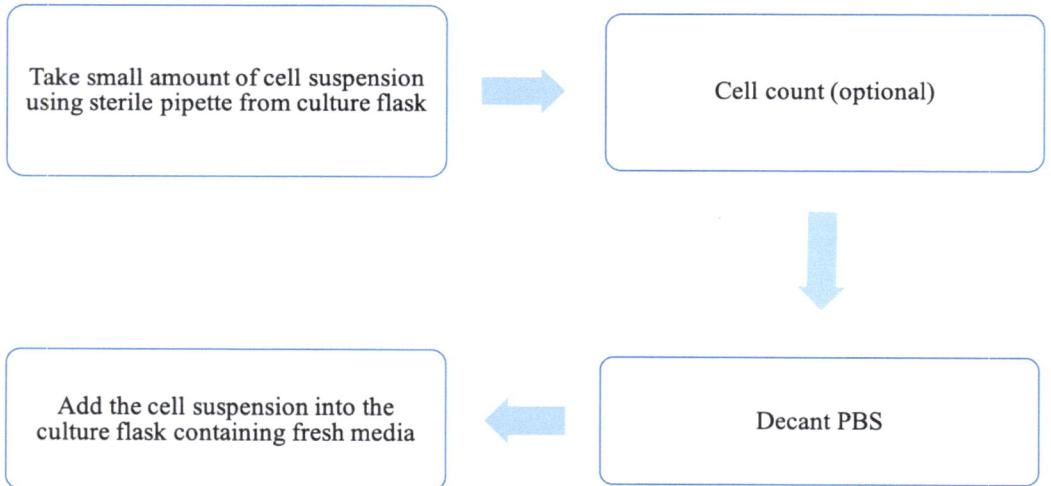

Take small amount of cell suspension using sterile pipette from culture flask → Cell count (optional)

Add the cell suspension into the culture flask containing fresh media ← Decant PBS

Fig. (4). Flow-diagram showing steps to follow in sub-culturing suspension cell cultures.

5. CRYOPRESERVATION OR FREEZING OF CELLS

Cryopreservation or freezing of cells is a vital procedure to develop a stock of cells. The cell lines are always prone to genetic drift, susceptible to contamination, and the finite cells are fated for senescence. As cell lines are valuable resources, so it is always necessary to freeze cells for long-term storage. The cryopreservation of cultured cells is done by storing them in cryovials in liquid nitrogen in a freezing mixture containing FBS/ FCS and cryoprotective agents such as dimethylsulfoxide (DMSO), and glycerol. The cryoprotective agents have lower cooling rate which reduces the risk of ice crystal formation that can damage the cells. It is recommended to handle DMSO using appropriate measures as it is toxic and known to facilitate the entry of organic molecules into tissues [3].

6. THAWING OF FROZEN CELLS

The thawing or reviving of frozen cells from a liquid nitrogen tank has to be done quickly to ensure a high proportion of cells to survive the procedure. The steps necessary to follow for thawing frozen cells are given below:

- Set the water bath at 37°C.
- Take the cryovial containing required cell cultures from the liquid nitrogen tank
- Thaw the frozen cells quickly (<1 minute) by keeping the cryovial in a floater in a 37°C water bath.
- Take 5-6 ml of pre-warm complete media in a 15 ml centrifuge tube in the biosafety cabinet/laminar hood under aseptic conditions.
- Add the thawed cells slowly to the pre-warm media.
- Centrifuge the tube at 200g for 5 minutes.
- Decant the supernatant to remove DMSO contained in the freezing mixture.
- Resuspend the cell pellet in pre-warm complete media.
- Add the cell suspension in the culture flask containing fresh complete media.
- Examine the culture flask under a microscope and keep it in a 37°C incubator humified with 5% CO_2.

7. CELL COUNTING

Cell counting is the routine procedure performed in cell culture experiments. Counting of cells and cell viability checks can be done easily and accurately by using automated cell counter. Alternatively, hemocytometer (or haemocytometer) with trypan blue dye exclusion assay can be used to count the total cells and check the viability percentage. Hemocytometer contains 4 major squares in 4 corners and each square contains 16 small squares. It is necessary to count all the small squares of 4 major squares. The steps necessary to follow for thawing frozen cells are given below:

- Wipe the glass hemocytometer and the coverslip with 70% ethanol.
- Add 10μl of cell suspension mixed with 10μl of Trypan blue dye (0.4%, w/v).
- Place the cover slip on the hemocytometer.
- Use a 10μl micropipette and transfer 10μl of cell suspension mixed with trypan blue to both the chambers of a hemocytometer.
- Carefully allow each chamber to fill by capillary action. Do not overfill or underfill the chambers.
- Carefully place the hemocytometer under microscope and count cells under 10X magnification.
- The viable cells remain unstained whereas; the non-viable or dead cells retain the stain and appear bluish in color.
- Each square of the hemocytometer represents the total volume of 0.1 mm^3 or 10^{-4}cm^3.
- As 1cm^3 is approximately 1ml. Therefore, total number of cells per ml=Average count per square x dilution factor x 10^4cells/ml.

$$\text{Cell viability (\%)} = \frac{\text{Total viable cells (unstained cells)}}{\text{Total cells (stained and unstained)}} \times 100$$

8. CELL VIABILITY AND CELL PROLIFERATION ASSAY BY USING MTT

Principle: Measuring cell viability and proliferation by using MTT (3-(4, 5-dimethylthiazolyl-2)-2,5-diphenyltetrazolium bromide) are the most important and routinely performed *in vitro* assays. The yellow tetrazolium MTT is reduced by metabolically active cells containing dehydrogenase enzymes and by reducing equivalents such as nicotinamide adenine dinucleotide (NADH) and nicotinamide adenine dinucleotide phosphate (NADPH). The resulting intracellular purple formazan can be solubilized by DMSO and the absorbance can be quantified spectrophotometrically at 570 nm. In the absence of cells, MTT reagent yields low absorbance [4]. The percentage of cell viability can be calculated by the following formula:

$$\% \text{ Cell viability} = \frac{\text{Absorbance of test well} - \text{Absorbance of blank}}{\text{Absorbance of control well} - \text{Absorbance of blank}} \times 100$$

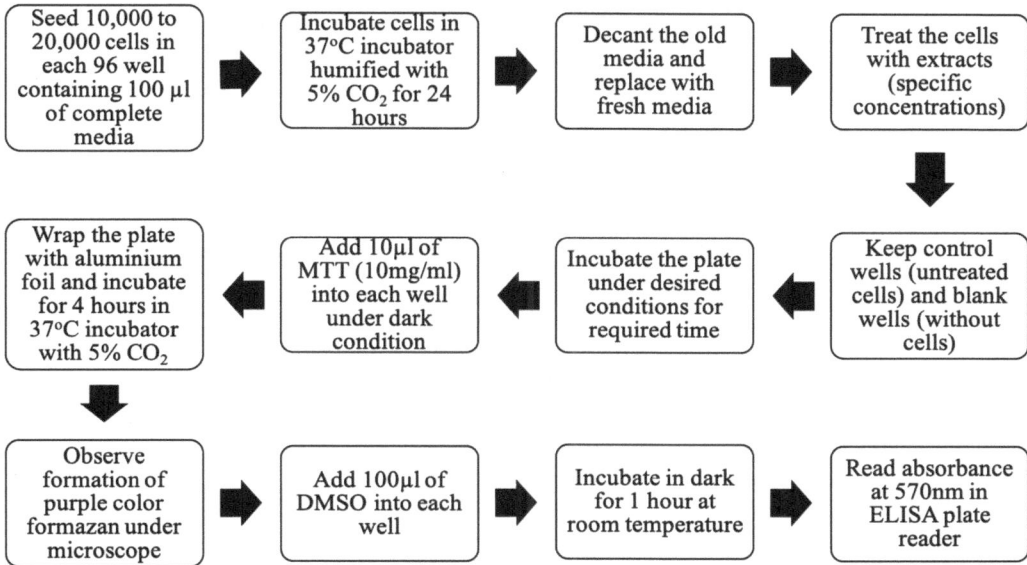

Seed 10,000 to 20,000 cells in each 96 well containing 100 μl of complete media → Incubate cells in 37°C incubator humified with 5% CO_2 for 24 hours → Decant the old media and replace with fresh media → Treat the cells with extracts (specific concentrations) ↓

Wrap the plate with aluminium foil and incubate for 4 hours in 37°C incubator with 5% CO_2 ← Add 10μl of MTT (10mg/ml) into each well under dark condition ← Incubate the plate under desired conditions for required time ← Keep control wells (untreated cells) and blank wells (without cells) ↓

Observe formation of purple color formazan under microscope → Add 100μl of DMSO into each well → Incubate in dark for 1 hour at room temperature → Read absorbance at 570nm in ELISA plate reader

Fig. (5). Flow-diagram showing steps in MTT assay.

9. CELL CULTURE CONTAMINANTS

The fundamental requirement of the cell culture laboratory is to protect cells from contamination. Cell culture contaminations lead to very serious consequences, if not controlled in the beginning. It is sometimes difficult to eliminate the contamination entirely, but it is possible to reduce the frequency of contamination by having a basic understanding of cell culture contaminations and by following stringent cell culture protocols [5, 6].

Cell culture contaminants can be broadly divided into two categories:

a. Chemical contaminants include impure culture media, sera, unsterilized water, presence of endotoxins and detergents.
b. Biological contaminants such as bacteria, molds, yeasts, viruses, mycoplasma and cross-contamination by other cell lines.

9.1. Overview of Biological Contaminants in Cell Culture

Primarily the biological contaminants in cell culture include bacterial contaminants, contaminants due to mold and virus, mycoplasma and yeast contaminations. It is necessary to identify the type of contamination to overshoot it while performing routine cell culture protocols.

9.1.1. Bacterial Contamination

Bacteria are ubiquitous and are the most commonly encountered biological contamination in cell culture. The bacterial contamination can be detected by naked eyes by simply visualizing the culture media. The infected media appear cloudy or turbid. The change in color of the infected culture media due to the drop in pH can also be detected. Under a microscope in low magnification, bacteria look like tiny, moving particles between the cells.

9.1.2. Mycoplasma Contamination

Mycoplasmas are the simplest and smallest self-replicating bacteria. They are very difficult to detect in cultures until they grow much to become highly dense. Mycoplasma infections can decrease the rate of cell proliferation, can alter the behavior, and metabolism of host cells. Mycoplasma contamination can cause agglutination in suspension cultures. They can persist in cell cultures without causing any cell death. Mycoplasma can be detected in cell cultures periodically by microbiological assays, ELISA, PCR, immunostaining, and fluorescent staining.

9.1.3. Mold Contamination

Molds belong to the kingdom Fungi, having multicellular filament- like structures known as hyphae and form a connected network of these filaments called mycelium. Under high power microscope, the mycelia_appear as thin_filament structures, and dense clumps of spores. These spores are resistant to adverse environments and they remain in dormant states. Only under suitable growth conditions, these spores become activated.

9.1.4. Yeast Contamination

Yeasts also belong to the kingdom Fungi, unicellular eukaryotic microorganisms. The cell cultures contaminated with yeasts become turbid and there is a change in pH of the culture once the contamination becomes dense. Under the microscope, yeasts appear to be individual particles either ovoidal or spherical in shape and may sometimes bud off to form smaller particles.

9.1.5. Virus Contamination

Viruses are microscopic infectious agents and they are very difficult to detect in culture owing to their extremely small size. Once encountered, it is very difficult to get rid of viral contamination. Moreover, virus-infected cell cultures can also pose a serious threat to the laboratory personnel. Viral infection can be detected by electron microscopy, enzyme-linked immunosorbent assays (ELISA), immunostaining, and PCR.

9.2. Cross Contamination

Cross-contamination in cell culture due to the presence of other cell lines, although not common, but may poses threat to the cell culture system. Obtaining cells from unrecognized cell banks/repositories, accidental inoculation of one cell line with another, accidental reuse of pipettes, and accidental mislabeling may lead to the problems of cross- contamination in the cell culture systems. The best practices to confirm the presence or absence of cross-contamination are DNA fingerprinting, karyotype analysis, and isotype analysis of the cell cultures [6].

The simple steps to follow for avoiding cross contaminations are:

- Obtain cells from reputed and recognized cell banks.
- Handle only one cell line in biosafety hood at one time.
- Clean work-place and empty the waste pots thoroughly after handling one cell line.
- Clear labeling of the reagent bottles and aliquots.
- Avoid sharing of reagent bottles between cell lines.

10. APPLICATIONS OF ANIMAL CELL CULTURE IN BIOMEDICAL RESEARCH

The cell culture techniques are widely used in various biomedical research areas such as in cell and molecular biology, virology, gene therapy, drug screening, nutritional studies *etc.* [1]. Table **2** shows some of the widely used cell lines in biomedical research. Some of the prominent research areas of animal cell culture are as follows:

- **Cancer research:** Animal cell cultures are widely used to study the cellular and biochemical changes occurring in the carcinogenesis process. Both the normal cells and cancer cells can be studied using animal cell culture techniques. Cancer cell lines of different types of cancers are commercially available in cell repositories. Novel anticancer drugs/compounds can be tested in the cancer cells before testing in the model animals.
- **Virology:** Animal cell cultures can be used as the model for replicating viruses for the production of vaccines instead of using animals. Animal cell cultures can be used to study the growth and developmental cycle of viruses.
- **Vaccine production:** Cell cultures can be used in the process of production of vaccines. The cultured cells are used in the production of viruses which in turn are used to produce vaccines. The vaccines for polio, chicken-pox, hepatitis B, and measles are produced using animal cell cultures.
- **Toxicity testing:** Animal cell cultures are used primarily for checking the effects of new drugs, compounds, plant crude extracts on the cells before testing them on the model animals.
- **Genetic engineering and gene therapy:** The cultured cells are used as the mode to introduce new genetic materials like DNA/RNA into cells in order to check gene expressions of new genes. Animal cells can be used genetically altered and can be used in gene therapy.
- **Drug screening and development:** Animal cell cultures are widely used in pharmaceutical companies to determine the efficacy and safe dosages of drugs. The cultured cells are widely used in screening lead drug candidates before being tested in model animals.

Table 2. Common cell lines routinely used in biomedical research.

Cell lines	Origin	Applications	Image
MIN6	Mouse (Pancreatic β cells)	Glucose stimulated insulin secretion; β cell regeneration	
HepG2	Human (Liver cancer cells)	Hepatocellular carcinoma	
HUVECs	Human (Endothelial cells)	Vascular ageing; endothelial dysfunction	-

(Table 2) cont.....

Cell lines	Origin	Applications	Image
Vero	Monkey (Kidney epithelial cells)	Verotoxin detection, study of malarial biology	
HEK293T	Human (Embryonic kidney cells)	Wound healing studies	
K562	Human Leukemia (Myeloid cells)	Chronic myelogenous leukemia	
HeLa	Human (Cervical cancer cells)	Cervical cancer research	-
PC12	Rat (pheochromocytoma of rat adrenal medulla)	Neuronal differentiation, brain diseases	-
U87-MG	Human (primary glioblastoma cells)	Brain cancer research	-
MCF7	Human (breast cancer cells)	Breast cancer research	-

CONCLUSION

Animal cell culture is an important investigational tool with robust applications in biomedical research and therapeutics. With the advancements in cell culture technologies, animal cell culture has become a fundamental basis for studying molecular, cellular and biochemical basis of various diseases such as cancer, diabetes. Animal cell culture techniques are routinely used for drug screening and testing, vaccine development, and have a prominent role in pharmaceutical industries. Thus, good aseptic techniques of animal cell cultures are essential for successful biomedical research.

CONSENT FOR PUBLICATION

Not applicable.

CONFLICT OF INTEREST

The author confirms that this chapter contents have no conflict of interest.

ACKNOWLEDGEMENTS

The authors extend their gratitude to Dr. Vinaykumar Rale, Director, Symbiosis School of Biological Sciences for providing constant support. The authors acknowledge Dr. Santosh Koratkar and Dr. Swagata Roy for providing the images of cells.

REFERENCES

[1]   Ryan JA. Introduction to animal cell culture. Technical Bulletin 2008.

[2]   Freshney RI. Basic principles of cell culture. Culture of cells for tissue engineering 2006 Feb; 9: 11-4.

[3]   Hay RJ. Preservation of cell-culture stocks in liquid nitrogen. TCA manual/Tissue Culture Association 1978 Jun 1; 4(2): 787-90.
[http://dx.doi.org/10.1007/BF00918397]

[4]   Mosmann T. Rapid colorimetric assay for cellular growth and survival: application to proliferation and cytotoxicity assays. J Immunol Methods 1983; 65(1-2): 55-63.
[http://dx.doi.org/10.1016/0022-1759(83)90303-4] [PMID: 6606682]

[5]   Drexler HG, Uphoff CC. Mycoplasma contamination of cell cultures: Incidence, sources, effects, detection, elimination, prevention. Cytotechnology 2002; 39(2): 75-90.
[http://dx.doi.org/10.1023/A:1022913015916] [PMID: 19003295]

[6]   Cabrera CM, Cobo F, Nieto A, *et al.* Identity tests: determination of cell line cross-contamination. Cytotechnology 2006; 51(2): 45-50.
[http://dx.doi.org/10.1007/s10616-006-9013-8] [PMID: 19002894]

CHAPTER 2

Real Time PCR as a Diagnostic Tool in Biomedical Sciences

Ritismita Devi[1], Rohit Saluja[2,3] and **Swapnil Sinha[1,*]**

[1] *Assam Downtown University, Panikhaiti, Guwahati-781026, Assam, India*

[2] *Department of Biochemistry, All India Institute of Medical Sciences AIIMS Bhopal, India*

[3] *Department of Biochemistry, All India Institute of Medical Sciences AIIMS Bibinagar, India*

Abstract: The present chapter highlights the various newer and advanced real-time quantitative PCR (qPCR) detection chemistries and their applications in biomedical research and molecular diagnostics. The chapter also summarizes the most advanced modification of the qPCR based technique along with next generation methods which have immense potential to revolutionize the field of biomedical sciences and molecular diagnostics.

Keywords: Molecular diagnostics, qPCR, SYBR green.

INTRODUCTION

In 1993, Higuchi *et al.* innovated a revolutionary refinement of PCR capable of detecting the amplification of nucleic acid chain in 'Real-Time', which included the use of intercalating dye ethidium bromide incorporated in a growing chain in each amplification cycle [1, 2]. Since its development, Quantitative Real-time PCR (qPCR) has gradually become an established method in laboratories for diagnostics and research applications. The ease of combining the polymerase chain reaction (PCR) with fluorescently labeled probes to monitor the incorporation of nucleotides in 'Real-Time' has imparted this procedure an excellent alternative to conventional end-point PCR. The qPCR technology is extremely sensitive, highly accurate with excellent reproducibility; moreover, the simultaneous 'assessment' of qPCR reactions in real-time abrogates subsequent steps of post-PCR monitoring and analysis (*e.g.* electrophoresis) and therefore reducing contamination and imparting accuracy.

For laboratory applications, qPCR technology has emerged as a hotspot in compe-

---
* **Corresponding author Swapnil Sinha:** Assam Downtown University, Panikhaiti, Guwahati-781026, Assam, India; Tel: +91 8011743775; E-mail: swaps.gene@gmail.com

Anupam Jyoti & Neetu Mishra (Eds.)

titive scientific market and was quickly commercialized by Applied Bio systems in 1996. Presently the qPCR instrumentation with numerous modifications customized for laboratory applications has been made available by leading companies including Bio-Rad, Qiagen, Roche Applied Science, Stratagene, *etc.* Real-time PCR has established its versatility in various research applications including differential gene expression analysis, microarray, genotyping, drug discovery and microRNA expression. qPCR has revolutionized the field of molecular diagnostics and its ever- expanding modifications are finding immense applications in biomedical fields like microbiology, veterinary sciences, agriculture, pharmacology, toxicology *etc.* More recently, the technique has provided new dimensions to high-throughput molecular diagnostics with fast, reliable and reproducible results.

This chapter offers a detailed description of various detection chemistries that are used in qPCR technique. Depending on the nature of the experiment and application, it is also explained which detection chemistry will suit best for the laboratory needs. The chapter also presents a comprehensive overview of the numerous advanced applications of the technique and how the technology can be utilized in future for the development and validation of drug discovery process.

Detection Chemistries

In their original work, Higuchi *et al.* used ethidium bromide (EtBr) dye as the detection molecule that non-specifically intercalates DNA [2]. EtBr and some other classical nucleic acid dyes often interfere with the PCR and therefore fluorescent dyes are almost exclusively used in qPCR now. qPCR detection probes can be divided into two classes, those labeled with non-specific dyes and other labeled probes which are sequence specific. Below is the detailed account of various detection chemistries used in real-time PCR:

Intercalating Dyes

These dyes bind to the minor groove of dsDNA and the dye molecules are incorporated in the growing DNA chain during PCR cycles. These dyes do not bind in a sequence specific manner and with each cycle, the number of molecules incorporated or accumulated in DNA amplicons increases which directly proportionates to the increase in fluorescence intensity with each cycle [3]. Since these dyes are non-sequence specific in nature, melt curve analysis (for the presence of non-specific products) and primer dimers should be performed during PCR process. Since during DNA denaturation, primer dimers and non-specific products melt (denature) at lower temperatures, they are presented as distinct peaks in melt curve plots [4].There are numerous commercial dyes available including SYBR Green [4], EvaGreen [5], SYTO [6], and BOXTO [7]. The most popular dye used in most of the qPCR applications is the SYBR Green dye mostly because of its very high DNA binding affinity [8, 9]. It is a cyanine dye possessing two positive charges during PCR [8] forming a DNA-dye complex which absorbs blue light (497nm) and emits green light (520nm). One advantage of using intercalating fluorescent dyes is their low cost as compared to labeled probes. EvaGreen is, another superior DNA binding dye which has enhanced resistance from PCR inhibitors and has improved efficiency and processivity [10]. The intercalating dyes are efficiently deployed in qPCR in applications like gene expression, pathogen detection [11], SNP genotyping [12], GMO detection [13] *etc.*

Fluorescent Labeled Oligonucleotide Probes

In many of the qPCR reactions, fluorophore labeled oligonucleotides called probes are deployed. At present, there are a variety of fluorophore-labeled probes offered by companies that are used in qPCR applications. Some popular ones are summarized below:

TaqMan Probes

They were first described by Holland *et al.* in 1991 [14]. Presently, they are marketed by Applied Biosystems [15, 16]. TaqMan probe assay is based on the 5'nuclease activity of *Taq*polymerase during PCR. Two TaqMan probes are designed for each polymorphic/target locus, one probe is complementary to the wild-type allele and the other to the mutant allele. Each probe has different dyes conjugated to their 5'OH ends along with a quencher at the 3'OH end. When the probes are not binding to the complementary DNA, the quencher quenches the fluorescence of the fluorophore. During qPCR, only specific complementary probe anneals to the template DNA and during extension step, *Taq*polymerase cleaves the 5' fluorescent dye resulting in the rise of its fluorescence Fig. (**1**). Mismatch probes, which do not anneal to the template are not cleaved by *Taq* polymerase imparting no fluorescence. The genotype of the sample is detected by measuring the fluorescence intensity of dyes used for each probe. The design and synthesis of TaqMan probes are relatively easy and the assays are fast, accurate and reproducible. TaqMan probes are widely used in SNP detection in both single and multiplex formats [17, 18]. They are also preferred assays for allelic discrimination, miRNA polymorphisms *etc.* [19, 20].

Molecular Beacons Probe

These were developed by Tyagi *et al.* in which they designed a single stranded hairpin oligonucleotide sequence of 20-30 bases complementary to the target DNA locus [21]. These probes have a fluorophore attached at its 5'-end of the ssDNAoligo which emits light of longer wavelength (lower energy). This energy is absorbed by a quencher attached at the 3'-end of the probe present in close proximity. The probe is designed to bind to the specific locus on target DNA and appearance of fluorescence signal works on FRET (Fluorescence Resonance Energy Transfer) principle. As depicted in Fig. (**2**), if the probe finds the specific complementary locus on the target DNA, the fluorophore at 5'-end will be separated from the quencher at 3'-end and hence the light emitted by the fluorophore will be recorded by the detector. If the probe doesn't bind to the target DNA, the light emitted by the fluorophore will be quenched by the quencher and there will be no fluorescence signal. Hairpin molecular beacons are used in a wide range of applications including SNP detection, Allele discrimination, pathogen detection, Viral load quantification, Gene expression analysis, *in vitro* quantification or detection *etc.* [22].

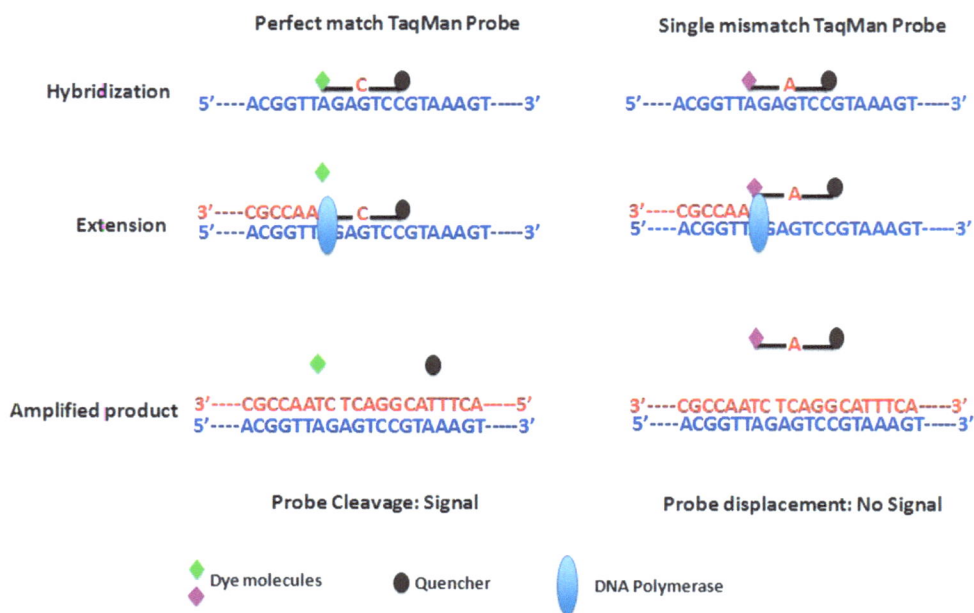

Perfect match TaqMan Probe

Single mismatch TaqMan Probe

Hybridization

5'----ACGGTTAGAGTCCGTAAAGT-----3'

5'----ACGGTTAGAGTCCGTAAAGT-----3'

Extension

3'----CGCCAA — C —
5'----ACGGTT GAGTCCGTAAAGT-----3'

3'----CGCCAA — A —
5'----ACGGTT GAGTCCGTAAAGT-----3'

Amplified product

3'----CGCCAATC TCAGGCATTTCA-----5'
5'----ACGGTTAGAGTCCGTAAAGT-----3'

3'----CGCCAATC TCAGGCATTTCA-----3'
5'----ACGGTTAGAGTCCGTAAAGT-----3'

Probe Cleavage: Signal

Probe displacement: No Signal

◆ Dye molecules   ● Quencher   DNA Polymerase

Fig. (1). Schematic representation of allelic discrimination achieved by *TaqMan* Probes.

Nucleic Acid Analogues

As the namesake, Nucleic acid analogues are compounds that are structurally similar to natural RNA and DNA. These analogues have modifications in phosphate backbones, pentose sugar or nitrogen bases [23] retaining all the properties of the native base with additional stability and enhanced affinity for target loci. A variety of nucleic acid analogues have been described in recent decade, *e.g.* 2'-O-methyl oligodeoxyribonucleotides or 2'-O-methyl RNA [24],

Peptide Nucleic Acids (PNAs) [25], Locked Nucleic Acids (LNAs) [26], Zip nucleic acids (ZNAs) [27] *etc.*, that are now extensively used in qPCR applications. These analogues, due to their improved features like resistance to nuclease degradation, have rapidly been incorporated in qPCR applications in place of natural oligonucleotides. Often, analogues are used in combination with unmodified DNA/RNA to enhance specificity and thermal stability. Nucleic acid analogues have been used for qPCR applications in cell- based assay [28], microarray [29], pathogen detection [30] *etc.*

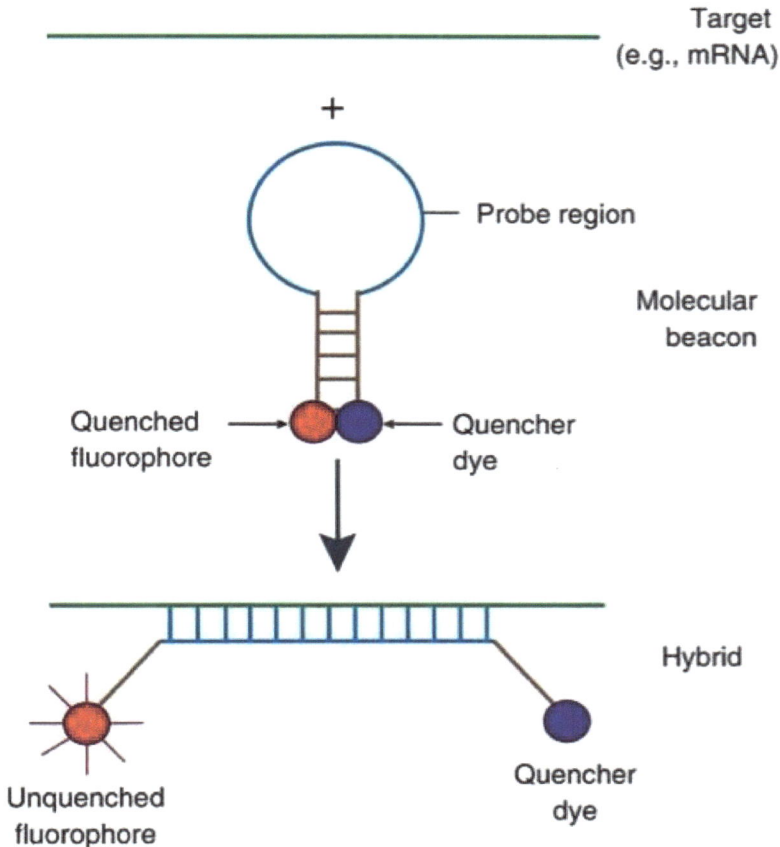

Fig. (2). Schematics of detection chemistry in Molecular Beacon.
Source: Nature Protocols volume 1, pages 1392–1398 (2006).

APPLICATIONS

Differential Gene Expression

One of the first areas where real-time PCR is primarily employed is to study the differential gene expression. Before the quantitative measurement of transcripts, gene expression detection was limited to a qualitative analysis by conventional PCR. Real-time PCR has proven to be the hallmark technique for gene expression based studies and has gained immense popularity due to its enhanced sensitivity, high accuracy and tremendous reproducibility. qPCR is routinely used in monitoring of differential gene expression in normal and diseased tissues/cells [31, 32], chromatin dynamics, coupling qPCR with Chromatin Immuno-precipitation (ChIP) [33, 34], plant gene expression [35] and in drug response [36, 37].

Cancer Diagnostics

Recently Wu *et al.* quantitatively detected E6 and E7 mRNAs using qPCR in cervical brushing cells from HPV patients overcoming limitations of low sensitivity and accuracy of traditional PCR methods [38]. This method can be used for cytological diagnosis of cervical neoplasia and can also help determine the severity of the lesions of viral infection. Deng *et al.* used qPCR assays to show the enhanced expression of Esophageal cancer-related gene ECRG4, which can be utilized as a biomarker in patients of gastric cancer [39].Various qPCR based assays have been developed to monitor the expression of miRNA in cell systems. Many companies (*e.g.* Qiagen, Sigma, Thermofisher *etc.*) have developed specific qPCR based miRNA assays that are now widely used for diagnosis and prognosis of many diseases including cancer. Recently, Gong *et al.* established miR-221 overexpression and its correlation with tumor stage, metastatic status, and response to chemotherapy pretreatment in Osteosarcoma patients [40]. At present, gastric cancer has no definitive biomarker or symptoms in its initial phases and is usually diagnosed in advanced phases. Recently, Abbasi *et al.* used real-time PCR to monitor the expression and methylation status of Reprimo gene in Gastric cancer patients [41]. Their study has shown promise that Reprimo gene assay could be developed as a diagnostic and prognostic biomarker for gastric cancer. Cell- free DNA found in human blood can be used as initial level biomarker for many diseases including cancer. However, low concentrations limit its detection by conventional PCR. Soliman and co-workers have used cell free microsatellite DNA for their quantitative estimation by real-time PCR in metastatic non-small lung cancer patients [42]. Recently, serum circulating microRNAs (c-miRNAs) served as useful biomarkers for cancer diagnosis. Fan *et al.* developed a qPCR based one-step branched rolling circle amplification (BRCA) method to estimate serum c-miRNAs levels for early diagnosis of breast cancer [43].

Bacterial Infection Diagnosis

Real-time PCR has found a very useful application in detection of pathogens and other microorganisms in biological fluids like blood and urine including tissue samples. qPCR has been established as a trusted method for detection of slow-growing, difficult-to-cultivate microbes and is fast replacing conventional microbiological techniques which are time-consuming and expensive. Many companies offer qPCR based pathogen detection kit with high sensitivity and a very low detection limit (*e.g.* Thermofisher Zika virus detection kit). Houmami *et al.* developed *K. kingae*-specific qPCR assay, targeting the malate dehydrogenase and used it for diagnosing *K. kingae* infections and carriage in 104 clinical specimens from children aged between 7 months and 7 years old [44]. Wang *et al.* developed a real-time reverse transcription-polymerase chain reaction (real-time RT-PCR) assay targeting two target genetic elements to detect HDV RNA and showed that the assay could be used for clinical detection of HDV infection in chronic HBV-infected patients [45]. Yan and co-workers used TaqMan probes for detection of *Mycobacterium leprae* DNA in paraffin-embedded skin biopsy specimens from leprosy and other dermatoses and bacterial DNA from 21 different species [46]. In another study, Ranjbar *et al.* evaluated the performance of TaqMan® probes to detect *Salmonella typhi*. TaqMan® real-time PCR assays were performed by pre-designed primers-probes for staG gene for detecting *S. typhi* [47].

Food Adulteration

Application of real-time PCR in detecting food borne pathogens and other DNA contamination has been instrumental in the food and diagnostic industry. Successful deployment of qPCR in food adulteration is primarily because of low detection limit and high specificity of measurement. Recently Schares *et al.* developed a Magnetic Capture-qPCR assay for quantifying *Toxoplasma gondii* infection in chicken samples [48]. In another method, Zhang *et al.* developed a rapid and sensitive qPCR based strategy coupled with microchip capillary electrophoresis for simultaneous detection of three foodborne pathogenic bacteria *viz. E. coli*, *S. aureus* and *S. typhinurium* [49].

Mutational Analysis for Predicting Predisposition to Disease(s)

Real-time PCR methods have been successfully used in SNP (single nucleotide polymorphisms) discovery and genotyping. Emerging qPCR based assays are now routinely being developed for genetic association studies including high-throughput genome-wide association studies [50]. There are various real-time PCR based methods being offered by companies for differentiating between the alternate alleles of polymorphism and also allele quantification. *TaqMan*SNP genotyping assays (Applied Biosystems) offer allele-specific fluorescent probes that can discriminate polymorphic alleles. BHQplus™ probes from Biosearch Technologies offer SNP genotyping probes which combines melt curve analysis tool with target hybridization stabilization during real-time PCR.

Advanced Techniques

Over the last decade, the advent of technologies like next generation sequencing, real-time PCR technique *etc* have undergone considerable modifications which are more sophisticated and customized to user needs. Automation of gene profiling methods and imparting it a high-throughput edge, have a huge potential to take expression-based studies to a broader range of clinical applications and expediting the drug discovery process. Combining ChIP with qPCR validation has contributed tremendously towards understanding the intricacies of cell mechanisms [51, 52]. Lim and co-workers have described an integrated 'Lab-o--chip' (LOC) microfluidic system, which combines both tissue sample preparation and multiplex real-time RT-PCR for multiplex gene expression analysis for point-of-care testing [53]. It can be semi-automated and is feasible for clinical lab setting. High-throughput automated qPCR based transcriptome analysis, quantitating the mRNA abundance has provided new dimensions to the personalized medicine research and has also contributed to the process of development of novel therapeutics [54].

CONCLUDING REMARKS

Even after the advent of various advanced NGS methods, qPCR is still considered as the 'gold standard' for gene expression profiling mostly because of its enhanced capability to detect low copy number transcripts. Recent advances are making successful attempts in coupling the high-throughput and cost-effective quality of NGS with fast, reliable and highly sensitive capability of qPCR to develop technologies that can be benchmark for lab-based basic research and drug discovery.

CONSENT FOR PUBLICATION

Not applicable.

CONFLICT OF INTEREST

The author confirms that this chapter contents have no conflict of interest.

ACKNOWLEDGEMENTS

This work was primarily supported by DBT, Government of India, Ref. No: BT/RLF/Re-entry/53/2013. RS acknowledges support from the Department of Biotechnology, India for Ramalingaswami Re-entry Fellowship.

REFERENCES

[1]   Higuchi R, Dollinger G, Walsh PS, Griffith R. Simultaneous amplification and detection of specific DNA sequences. Biotechnology (N Y) 1992; 10(4): 413-7.
        [http://dx.doi.org/10.1038/nbt0492-413] [PMID: 1368485]

[2]   Higuchi R, Fockler C, Dollinger G, Watson R. Kinetic PCR analysis: real-time monitoring of DNA amplification reactions. Biotechnology (N Y) 1993; 11(9): 1026-30.
        [PMID: 7764001]

[3]   Wittwer CT, Herrmann MG, Moss AA, Rasmussen RP. Continuous fluorescence monitoring of rapid cycle DNA amplification. Biotechniques 1997; 22(1): 130-131, 134-138.
        [http://dx.doi.org/10.2144/97221bi01] [PMID: 8994660]

[4]   Chou Q, Russell M, Birch DE, Raymond J, Bloch W. Prevention of pre-PCR mis-priming and primer dimerization improves low-copy-number amplifications. Nucleic Acids Res 1992; 20(7): 1717-23.
        [http://dx.doi.org/10.1093/nar/20.7.1717] [PMID: 1579465]

[5]   Wang W, Chen K, Xu C. DNA quantification using EvaGreen and a real-time PCR instrument. Anal Biochem 2006; 356(2): 303-5.
        [http://dx.doi.org/10.1016/j.ab.2006.05.027] [PMID: 16797474]

[6]   Gudnason H, Dufva M, Bang DD, Wolff A. Comparison of multiple DNA dyes for real-time PCR: effects of dye concentration and sequence composition on DNA amplification and melting temperature. Nucleic Acids Res 2007; 35(19)e127
        [http://dx.doi.org/10.1093/nar/gkm671] [PMID: 17897966]

[7]   Bengtsson M, Karlsson HJ, Westman G, Kubista M. A new minor groove binding asymmetric cyanine reporter dye for real-time PCR. Nucleic Acids Res 2003; 31(8)e45
        [http://dx.doi.org/10.1093/nar/gng045] [PMID: 12682380]

[8]   Morrison TB, Weis JJ, Wittwer CT. Quantification of low-copy transcripts by continuous SYBR Green I monitoring during amplification. Biotechniques 1998; 24(6): 954-958, 960, 962.
        [PMID: 9631186]

[9]   Zipper H, Brunner H, Bernhagen J, Vitzthum F. Investigations on DNA intercalation and surface binding by SYBR Green I, its structure determination and methodological implications. Nucleic Acids Res 2004; 32(12)e103
        [http://dx.doi.org/10.1093/nar/gnh101] [PMID: 15249599]

[10]   Mao F, Leung W-Y, Xin X. Characterization of EvaGreen and the implication of its physicochemical properties for qPCR applications. BMC Biotechnol 2007; 7: 76.
        [http://dx.doi.org/10.1186/1472-6750-7-76] [PMID: 17996102]

[11]   He P, Chen Z, Luo J, *et al*. Multiplex real-time PCR assay for detection of pathogenic Vibrio parahaemolyticus strains. Mol Cell Probes 2014; 28(5-6): 246-50.
        [http://dx.doi.org/10.1016/j.mcp.2014.06.001] [PMID: 24924797]

[12]   Li YD, Chu ZZ, Liu XG *et al*. A cost-effective high-resolution melting approach using the EvaGreen dye for DNA polymorphism detection and genotypingin plants. J Integr Plant Biol 2010; 52(12): 1036-42.

[13]   Akiyama H, Nakamura F, Yamada C, *et al*. A screening method for the detection of the 35S promoter and the nopaline synthase terminator in genetically modified organisms in a real-time multiplex polymerase chain reaction using high-resolution melting-curve analysis. Biol Pharm Bull 2009; 32(11): 1824-9.
        [http://dx.doi.org/10.1248/bpb.32.1824] [PMID: 19881291]

[14]   Holland PM, Abramson RD, Watson R, Gelfand DH. Detection of specific polymerase chain reaction product by utilizing the 5'----3' exonuclease activity of Thermus aquaticus DNA polymerase. Proc Natl Acad Sci USA 1991; 88(16): 7276-80.
        [http://dx.doi.org/10.1073/pnas.88.16.7276] [PMID: 1871133]

[15]   Gibson UE, Heid CA, Williams PM. A novel method for real time quantitative RT-PCR. Genome Res 1996; 6(10): 995-1001.
[http://dx.doi.org/10.1101/gr.6.10.995] [PMID: 8908519]

[16]   Clegg RM. Fluorescence resonance energy transfer and nucleic acids. Methods Enzymol 1992; 211: 353-88.
[http://dx.doi.org/10.1016/0076-6879(92)11020-J] [PMID: 1406315]

[17]   Woodward J. Bi-allelic SNP genotyping using the TaqMan® assay. Methods Mol Biol 2014; 1145: 67-74.
[http://dx.doi.org/10.1007/978-1-4939-0446-4_6] [PMID: 24816660]

[18]   Li HM, Zhang TP, Li XM, Pan HF, Ma DC. Association of single nucleotide polymorphisms in resisting gene with rheumatoid arthritis in a Chinese population. J Clin Lab Anal 2018; 32(9)e22595
[http://dx.doi.org/10.1002/jcla.22595] [PMID: 29978502]

[19]   Gravel A, Sinnett D, Flamand L. Frequency of chromosomally-integrated human herpesvirus 6 in children with acute lymphoblastic leukemia. PLoS One 2013; 8(12)e84322
[http://dx.doi.org/10.1371/journal.pone.0084322] [PMID: 24386368]

[20]   Pi L, Fu L, Xu Y, *et al.* Lack of association between miR-218 rs11134527 A>G and Kawasaki disease susceptibility. Biosci Rep 2018; 38(3)

[21]   Tyagi S, Kramer FR. Molecular beacons: probes that fluoresce upon hybridization. Nat Biotechnol 1996; 14(3): 303-8.
[http://dx.doi.org/10.1038/nbt0396-303] [PMID: 9630890]

[22]   Vet JA, Van der Rijt BJ, Blom HJ. Molecular beacons: colorful analysis of nucleic acids. Expert Rev Mol Diagn 2002; 2(1): 77-86.
[http://dx.doi.org/10.1586/14737159.2.1.77] [PMID: 11963813]

[23]   Petersson B, Nielsen BB, Rasmussen H, *et al.* Crystal structure of a partly self-complementary peptide nucleic acid (PNA) oligomer showing a duplex-triplex network. J Am Chem Soc 2005; 127(5): 1424-30.
[http://dx.doi.org/10.1021/ja0458726] [PMID: 15686374]

[24]   Inoue H, Hayase Y, Imura A, Iwai S, Miura K, Ohtsuka E. Synthesis and hybridization studies on two complementary nona(2′-O-methyl)ribonucleotides. Nucleic Acids Res 1987; 15(15): 6131-48.
[http://dx.doi.org/10.1093/nar/15.15.6131] [PMID: 3627981]

[25]   Nielsen PE, Egholm M, Berg RH, Buchardt O. Sequence-selective recognition of DNA by strand displacement with a thymine-substituted polyamide. Science 1991; 254(5037): 1497-500.
[http://dx.doi.org/10.1126/science.1962210] [PMID: 1962210]

[26]   Singh SK, Nielsen P, Koshkin AA, *et al.* LNA (locked nucleic acids): Synthesis and high-affinity nucleic acid recognition. Chem Commun (Camb) 1998; 4: 455-6.
[http://dx.doi.org/10.1039/a708608c]

[27]   Voirin E, Behr JP, Kotera M. Versatile synthesis of oligodeoxyribonucleotide-oligospermine conjugates. Nat Protoc 2007; 2(6): 1360-7.
[http://dx.doi.org/10.1038/nprot.2007.177] [PMID: 17545974]

[28]   Rezaei F, Daryani A, Sharifi M, *et al.* miR-20 a inhibition using locked nucleic acid (LNA) technology and its effects on apoptosis of human macrophages infected by Toxoplasma gondii RH strain. Microb Pathog 2018; 121: 269-76.
[http://dx.doi.org/10.1016/j.micpath.2018.05.030] [PMID: 29800695]

[29]   Henihan G, Schulze H, Corrigan DK, *et al.* Label- and amplification-free electrochemical detection of bacterial ribosomal RNA. Biosens Bioelectron 2016; 81: 487-94.
[http://dx.doi.org/10.1016/j.bios.2016.03.037] [PMID: 27016627]

[30]   Iranmanesh Z, Mollaie HR, Arabzadeh SA, Zahedi MJ, Fazlalipour M, Ebrahimi S. Evaluation of the

frequency of the IL-28 polymorphism (rs8099917) in patients with chronic hepatitis C using Zip nucleic acid probes, Kerman, Southeast of Iran. Asian Pac J Cancer Prev 2015; 16(5): 1919-24.
[http://dx.doi.org/10.7314/APJCP.2015.16.5.1919] [PMID: 25773845]

[31]   Ding C, Cantor CR. Quantitative analysis of nucleic acids--the last few years of progress. J Biochem Mol Biol 2004; 37(1): 1-10.
[PMID: 14761298]

[32]   Alvaro A, Solà R, Rosales R, *et al.* Gene expression analysis of a human enterocyte cell line reveals downregulation of cholesterol biosynthesis in response to short-chain fatty acids. IUBMB Life 2008; 60(11): 757-64.
[http://dx.doi.org/10.1002/iub.110] [PMID: 18642346]

[33]   Mendoza MA, Panizza S, Klein F. Analysis of protein-DNA interactions during meiosis by quantitative chromatin immunoprecipitation (qChIP). Methods Mol Biol 2009; 557: 267-83.
[http://dx.doi.org/10.1007/978-1-59745-527-5_17] [PMID: 19799188]

[34]   Li G, Dzilic E, Flores N, Shieh A, Wu SM. Strategies for the acquisition of transcriptional and epigenetic information in single cells. J Thorac Dis 2017; 9 (Suppl. 1): S9-S16.
[http://dx.doi.org/10.21037/jtd.2016.08.17] [PMID: 28446964]

[35]   Zhang J, Sun XL, Zhang LG, Hui MX, Zhang MK. Analysis of differential gene expression during floral bud abortion in radish (Raphanus sativus L.). Genet Mol Res 2013; 12(3): 2507-16.
[http://dx.doi.org/10.4238/2013.July.24.5] [PMID: 23979885]

[36]   Bonilla E, del Mazo J. Deregulation of gene expression in fetal oocytes exposed to doxorubicin. J Biochem Pharmacol 2003; 63: 1701-7.

[37]   Ono A, Hirooka K, Nakano Y, Nitta E, Nishiyama A, Tsujikawa A. Gene expression changes in the retina after systemic administration of aldosterone. Jpn J Ophthalmol 2018; 62(4): 499-507.
[http://dx.doi.org/10.1007/s10384-018-0595-4] [PMID: 29713904]

[38]   Wu MZ, Li WN, Cha N, *et al.* Diagnostic Utility of HPV16 E6 mRNA or E7 mRNA Quantitative Expression for Cervical Cells of Patients with Dysplasia and Carcinoma. Cell Transplant 2018; 27(9): 1401-6.
[http://dx.doi.org/10.1177/0963689718788521] [PMID: 30056761]

[39]   Deng P, Chang XJ, Gao ZM, *et al.* Down regulation and DNA methylation of ECRG4 in gastric cancer. Onco Targets Ther 2018; 11: 4019-28.
[http://dx.doi.org/10.2147/OTT.S161200] [PMID: 30034241]

[40]   Gong N, Gong M. MiRNA-221 from tissue may predict the prognosis of patients with osteosarcoma. Medicine (Baltimore) 2018; 97(29)e11100
[http://dx.doi.org/10.1097/MD.0000000000011100] [PMID: 30024497]

[41]   Abbasi A, Heydari S. Studying the expression rate and methylation of Reprimo gene in the blood of patients suffering from gastric cancer. Eur J Transl Myol 2018; 28(2): 7423.
[http://dx.doi.org/10.4081/ejtm.2018.7423] [PMID: 29991989]

[42]   Soliman SE, Alhanafy AM, Habib MSE, Hagag M, Ibrahem RAL. Serum circulating cell free DNA as potential diagnostic and prognostic biomarker in non small cell lung cancer. Biochem Biophys Rep 2018; 15: 45-51.
[http://dx.doi.org/10.1016/j.bbrep.2018.06.002] [PMID: 29984326]

[43]   Fan T, Mao Y, Sun Q, *et al.* Branched rolling circle amplification method for measuring serum circulating microRNA levels for early breast cancer detection. Cancer Sci 2018; 109(9): 2897-906.
[http://dx.doi.org/10.1111/cas.13725] [PMID: 29981251]

[44]   El Houmami N, Durand GA, Bzdrenga J, *et al.* A new highly sensitive and specific real-time PCR assay targeting the malate dehydrogenase gene of Kingellakingae and application to 201 pediatric clinical specimens. J Clin Microbiol 2018; 56(8): 56.
[http://dx.doi.org/10.1128/JCM.00505-18] [PMID: 29875189]

[45]   Wang Y, Glenn JS, Winters MA, *et al.* A new dual-targeting real-time RT-PCR assay for hepatitis D virus RNA detection. Diagn Microbiol Infect Dis 2018; 30173-1.

[46]   Yan W, Xing Y, Yuan LC, *et al.* Application of RLEP real-time PCR for detection of M. leprae DNA in paraffin-embedded skin biopsy specimens for diagnosis of paucibacillary leprosy. Am J Trop Med Hyg 2014; 90(3): 524-9.
[http://dx.doi.org/10.4269/ajtmh.13-0659] [PMID: 24493677]

[47]   Ranjbar R, Naghoni A, Farshad S, *et al.* Use of TaqMan® real-time PCR for rapid detection of Salmonella enterica serovar Typhi. Acta Microbiol Immunol Hung 2014; 61(2): 121-30.
[http://dx.doi.org/10.1556/AMicr.61.2014.2.3] [PMID: 24939681]

[48]   Schares G, Koethe M, Bangoura B, *et al.* Toxoplasma gondii infections in chickens - performance of various antibody detection techniques in serum and meat juice relative to bioassay and DNA detection methods. Int J Parasitol 2018; 48(9-10): 751-62.
[http://dx.doi.org/10.1016/j.ijpara.2018.03.007] [PMID: 29782830]

[49]   Zhang Y, Zhu L, Zhang Y, He P, Wang Q. Simultaneous detection of three foodborne pathogenic bacteria in food samples by microchip capillary electrophoresis in combination with polymerase chain reaction. J Chromatogr A 2018; 1555: 100-5.
[http://dx.doi.org/10.1016/j.chroma.2018.04.058] [PMID: 29724645]

[50]   Matsuda K. PCR-Based Detection Methods for Single-Nucleotide Polymorphism or Mutation: Real-Time PCR and Its Substantial Contribution Toward Technological Refinement. Adv Clin Chem 2017; 80: 45-72.
[http://dx.doi.org/10.1016/bs.acc.2016.11.002] [PMID: 28431642]

[51]   Song Z, Ren H, Gao S, Zhao X, Zhang H, Hao J. The clinical significance and regulation mechanism of hypoxia-inducible factor-1 and miR-191 expression in pancreatic cancer. Tumour Biol 2014; 35(11): 11319-28.
[http://dx.doi.org/10.1007/s13277-014-2452-5] [PMID: 25119596]

[52]   Zimmers SM, Browne EP, Williams KE, *et al.* TROP2 methylation and expression in tamoxifen-resistant breast cancer. Cancer Cell Int 2018; 18: 94.
[http://dx.doi.org/10.1186/s12935-018-0589-9] [PMID: 30002602]

[53]   Lim GS, Chang JS, Lei Z, *et al.* A lab-on-a-chip system integrating tissue sample preparation and multiplex RT-qPCR for gene expression analysis in point-of-care hepatotoxicity assessment. Lab Chip 2015; 15(20): 4032-43.
[http://dx.doi.org/10.1039/C5LC00798D] [PMID: 26329655]

[54]   Kabir MF, Mohd Ali J, Haji Hashim O. Microarray gene expression profiling in colorectal (HCT116) and hepatocellular (HepG2) carcinoma cell lines treated with *Melicope ptelefolia* leaf extract reveals transcriptome profiles exhibiting anticancer activity. Peer J 2018; 6e5203
[http://dx.doi.org/10.7717/peerj.5203] [PMID: 30042885]

CHAPTER 3

Flow Cytometry: Basics and Applications

Juhi Saxena[1] and **Anupam Jyoti**[2,*]

[1] *Dr. B. Lal Institute of Biotechnology, 6-E, Malviya Industrial Area, Jaipur 302017, India*

[2] *Amity Institute of Biotechnology, Amity University Rajasthan, Amity Education Valley, Kant Kalwar, NH-11C, Jaipur-Delhi Highway, Jaipur, India*

Abstract: Flow cytometer is a sophisticated and analytical tool which analyzes the phenotypic and functional characteristics of a cell in much lesser time. It measures the light scattering features, mainly forward scatter and side scatter properties of the cell. Optics, fluidics, and electronics are the main elements flow cytometer. Hydrodynamic focusing of the sheath fluid present in a sheath chamber enables the streamline motion of cells. This allows the encounter of each cell separately with the laser. A photodiode, photomultiplier tubes, optical filters, and beam splitters collect and emit the light or fluorescence in the form of signals. The signals are further processed by the electronic component and with the help of software, data are analyzed. For a cell to emit fluorescence, a number of fluorochromes (FITC, Alexa Fluor 488, Alexa Fluor 647 *etc.*) and dyes (propidium iodide, Hoechst, ethidium bromide *etc.*) are available. These fluorochromes may be conjugated with antibodies and emit fluorescence when excited by the laser. A diverse range of applications are possible using flow cytometer like apoptosis assay, cell cycle analysis, free radical generation assay, cytokine estimation, immunophenotyping, phagocytosis assay and many more. An advanced application of flow cytometer receiving attention is cell sorting. Individual cells or cell types can be sorted from a mixture of cells using physical fluorescence emission features of the cells named fluorescence assisted cell sorter.

Keywords: Apoptosis, Cell cycle, Cells sorting, Fluorochromes, Fluorescence, Immunophenotyping, Laser, Light scattering, Photomultiplier tube.

INTRODUCTION

Flow cytometry is defined as the measurement of cells under flow condition. It is a powerful tool for studying the functional features and phenotypes of the cells. In flow cytometer cells are passed through a light beam or laser and scatter the light in all directions. The cells are discriminated on the basis of the light scattered in forward direction (forward scattering or FSC) and sides (side scattering or SSC).

---

* **Corresponding author Dr Anupam Jyoti:** Amity Institute of Biotechnology, Amity University Rajasthan, Amity Education Valley, Kant Kalwar, NH-11C, Jaipur-Delhi Highway, Jaipur, India; Tel: 01426 405 678; E-mail: ajyoti@jpr.amity.edu

FSC is the amount of light that cells scatter at low angles (0.5–10°) from the axis, whereas light scattered in the side *i.e.* 90° of the incident light is called SSC (Fig. **1**). FSC is directly related to the size of the cell whereas SSC is directly related to the complexity or granularity of the cells [1].

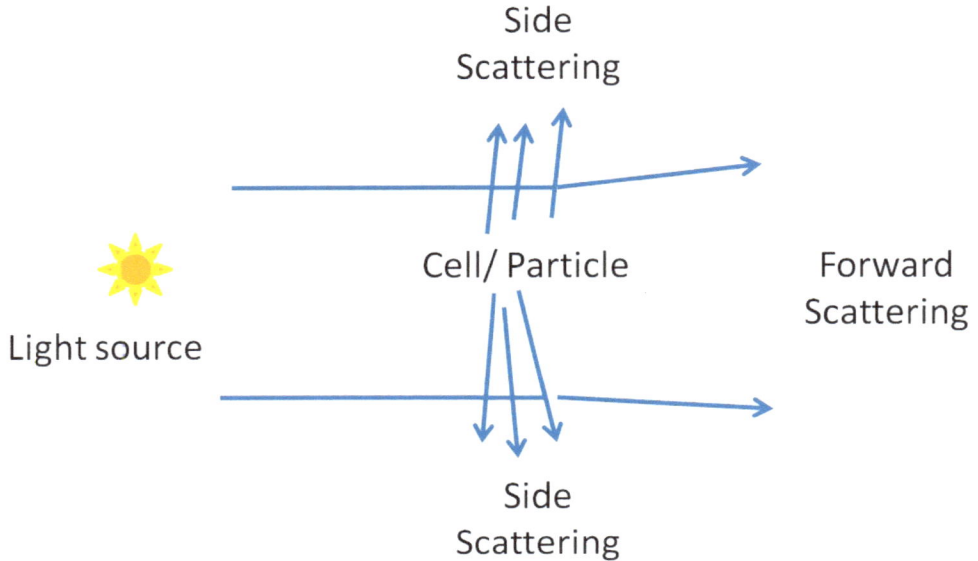

Fig. (1). FSC and SSC.

The cells loaded in the flow cytometer must be labeled with fluorescent tags including labeled antibody or organelle-specific dye or product binding dye. This is facilitated in the identification of cell types, estimation of DNA content and various cellular functional assays including free radical generation, apoptosis assay, phagocytosis assay and many more.

Components of a Flow Cytometer

The components of the flow cytometer can be categorized into fluidics, optics and electronic sections.

Fluidics

Fluidics is an integral component of a typical flow cytometer which facilitates the transport of cells or particles to the laser beam. The fluid containing the sample as well as sheath fluid is driven by the pressure supplied by the compressor (Fig. **2**).

When cell suspension is injected into a flow cytometer, the cells get dispersed randomly throughout the available free space. In order to face the laser, beam cells are aligned in a single streamline. Typically, the fluidic system contains a core where the sample is injected and flows. This is surrounded by sheath fluid in the outer layer which flows faster than the sample flow. While moving the sheath fluid, it imparts a drag effect on the core. Flowing fluid results in the narrowing of the sample containing core fluid. This results in high velocity of the core fluid at the centre and low velocity at the wall makes a parabolic movement [2]. Due to this, cells or particles move in a single file, a process known as hydrodynamic focusing (Fig. **3**). During this whole process, the core fluid containing sample is not mixed with the sheath fluid due to laminar flow.

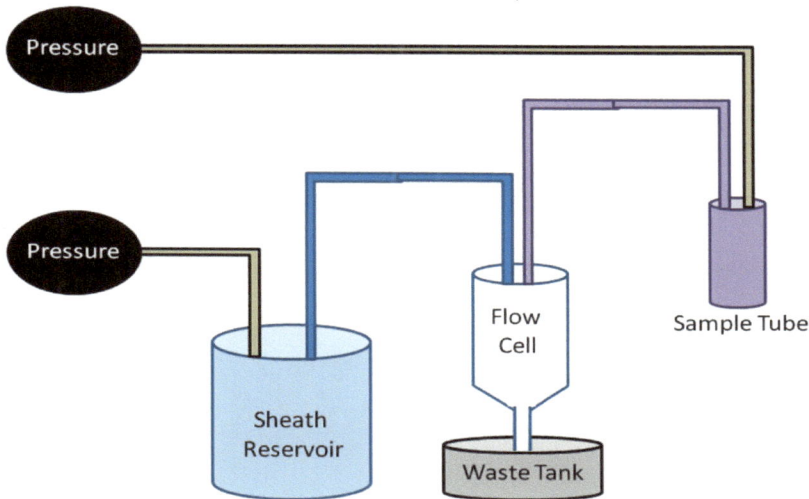

Fig. (2). Fluidics of a flow cytometer.

To achieve this, the sample pressure must be higher as compared to the sheath pressure. High sample pressure increases the width of the sample core hence it increases the flow rate which results in the entry of more cells in the stream. Due to this, some cells do not cross the laser beam in the centre. On the other hand, a slow flow rate decreases the width of the sample core and converges the cells to the narrow width of the sample core. This allows maximum cells to pass through the centre of the laser beam.

Optics

The optical system consists of a laser, photodiode, photomultiplier tube (PMT), filters and beam splitters. A typical optical layout of the flow cytometer is shown (Fig. **4**).

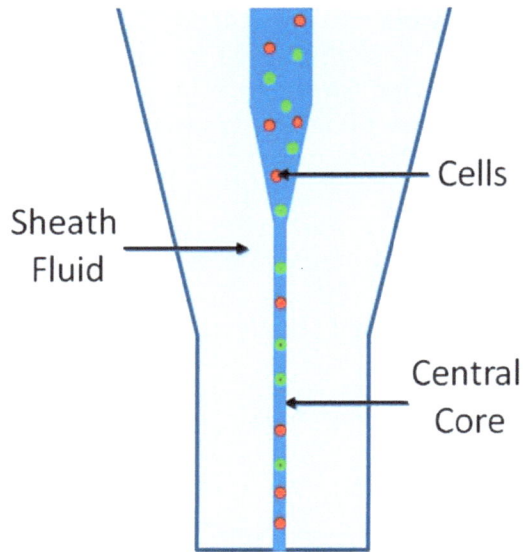

Fig. (3). Hydrodynamic focusing.

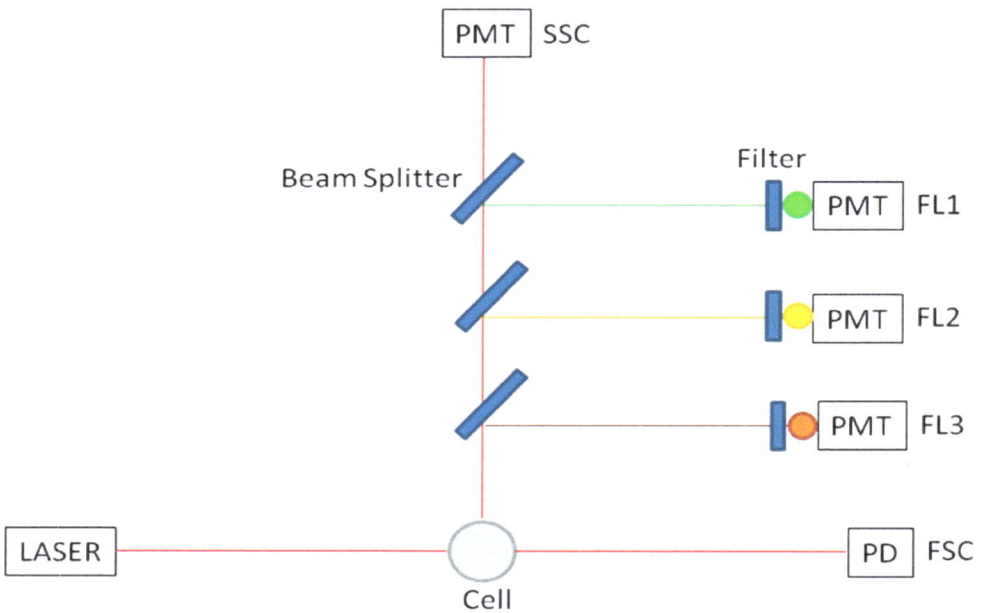

Fig. (4). Optical layout of a flow cytometer.

Laser – Laser stands for light amplification by stimulated emission of radiation. A laser is defined as a source that emits coherent light. Lasers are a preferred light source in a flow cytometer because they produce a high-intensity beam of a single colour (monochromatic) light. Some of the commonly used lasers in flow cytometer are argon ion, krypton ion, and helium/neon.

Photodiode - Photodiode is a semiconductor device that converts photon of light into an electron of the current. When photons are encountered with a photodiode, electric current is generated. It is used to capture the light scattered in the forward direction of a cell or particle in a flow cytometer.

PMTs – PMTs consist of a series of dynodes aligned serially in a tube. When the photons are encountered with PMTs, a series of electrons are generated in an amplified form. The photons generated from the reduced side-scattered light produce a large number of electrons after being encountered with a series of dynodes in PMTs (Fig. **5**).

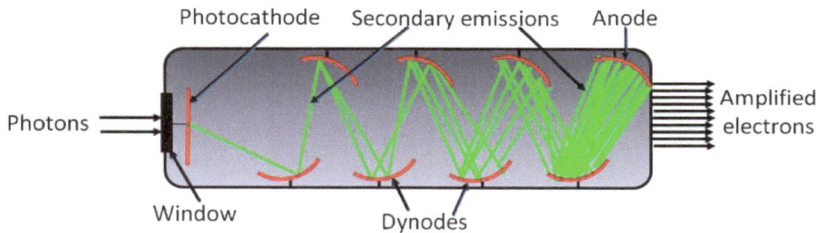

Fig. (5). PMT.

Optical Filters - The optical filters select the correct wavelength of light to pass through. These filters retain a fixed wavelength of light and allow others to pass. Three major filter types including long pass, short pass and bandpass are the most common. Long pass filters permit the light to pass over a certain wavelength. Short pass filters allow the light to pass beneath a certain wavelength. A certain wavelength range of light is passed through bandpass filters (Fig. **4**). All these filters block light by absorption.

Beam Splitters - These are metallic-coated quartz substrates and are designed to work at a 45° angle of incidence. Numbers indicate reflection/transmission values.

Electronics

The signal obtained after PMTs is further processed to get it in a readable format. The setting of the instrument is fixed, therefore, it can discriminate the signals generated from the desired particle or cell with signals generated from debris.

Pulse Processor – When the particle or cell encounters a laser beam, it scatters the light or emits fluorescence in the form of photons. These photons strike the PMT or photodiode and convert into the electrons and generate electric current. This electric current is further converted into the voltage pulse by means of an amplifier. This voltage pulse is typically a bell-shaped curve. The formation and shape of the curve are dependent on the movement of the particle across the beam. When the particle enters the beam voltage, pulse formation starts, at the centre of the beam the highest point of the curve is formed and at the exit of the beam, complete pulse is formed.

Analog-to-Digital Converter (ADC) – Once the voltage pulse is generated, it is converted into a digital system by means of ADC. The digitized signal corresponding to each voltage pulse is displayed on the data sheet and saved into the system memory.

Fluorochromes

Fluorochromes can be defined as chemicals that absorb light at a particular wavelength and further emit it at a different wavelength [3]. The wavelength on which light is absorbed is called excitation wavelength and released light is called emission wavelength. Basically, when a fluorochrome absorbs light of a certain wavelength, its electrons become excited and jump to a higher energy orbital. This step is called excitation and it lasts only a few nanoseconds. Furthermore, the electrons return to their lower energy orbital and emit light of different wavelengths. This step is called emission and is dependent upon the internal conformational change in the fluorochrome. During this process, some of the energy is dissipated in the form of heat, hence less energy is released. Therefore, emission fluorescence is of higher wavelength as compared to the exciting light wavelength [4, 5].

Maximal Excitation and Maximal Emission

The wavelength on which a fluorochrome absorbs light is an important parameter which influences the intensity of emission fluorescence. This can be understood by taking an example of FITC (fluorescein isothiocyanate). FITC absorbs light ranging from 400 to 550 nm. FITC absorbs a maximum number of photons at 490 nm, therefore, it is called λ_{max} for excitation. This lies in the blue region of the spectra hence blue Argon-ion laser is commonly used to excite FITC. Similarly, the emission range of FITC falls from 475 to 700 nm. However, maximum emission is at 525 nm which lies in the green spectrum. This is called λ_{max} for emission. In flow cytometer, different filters are used to screen out the different emitting light except that of λ_{max} for emission, hence FITC is visible as a green colour.

Fluorescence Compensation

It is an important fact when studying two or more different fluorochromes in a single study. The emission profile or spectra of fluorochromes will overlap which does not allow true measurement of fluorescence emitted by each fluorochrome. However, if we take two fluorochromes whose emission spectra are widespread in such a way that their emission profile will not coincide, such a problem can be avoided, but this is not always possible.

Fluorescence compensation is the solution that determines the amount of fluorescence a fluorochrome will have in a channel that was not assigned specifically to measure it. Flow cytometers are loaded with the compensation software which compensates automatically using simple mathematics.

Illustration 1: In an experiment, two different fluorochromes; say FITC and PE were used. FITC and PE are detected in FL1 and FL2 channels respectively. However, 10% of FITC fell into FL2 channel and 5% of PE fell in the FL1 channel. This is due to the fact that both FITC and PE have a closer emission profile. True reading for FITC = (Fluorescence measured in FL1) - (5% of PE's fluorescence)

True reading for PE = (Fluorescence measured in FL2) - (10% of FITC's fluorescence)

Data Analysis

There are two ways to display the data on the computer after acquisition – dot plot and histogram.

Dot Plot – In dot plot, as the name suggests, dots are displayed between x-axis and y-axis. Each dot represents a single cell or particle. Usually, these dots are distributed according to their FSC or SSC or fluorescence. More homogeneous the mixture of cell denser will be the distribution of dots in the plot and *vice versa* (Fig. **6**).

Histograms – Histograms are the graphs that are displayed as a single parameter (fluorescence) on the horizontal axis and the number of cells on the vertical axis. Ideally, the peak of the histogram should be narrow, however, in some of the cases the histogram peak broadens due to heterogeneity in the population. It is always advisable to select the least scattered or denser population of cells for the analysis. This type of gating gives a true picture of analysis for which cells were selected (Fig. **7a**). On the other hand, if two distinct populations of cells have been selected for the parameter (fluorescence) analysis, we will find two different histograms (Fig. **7b**). The distance between both the histograms depends upon how these distinct populations of cells are behaving for a parameter (fluorescence).

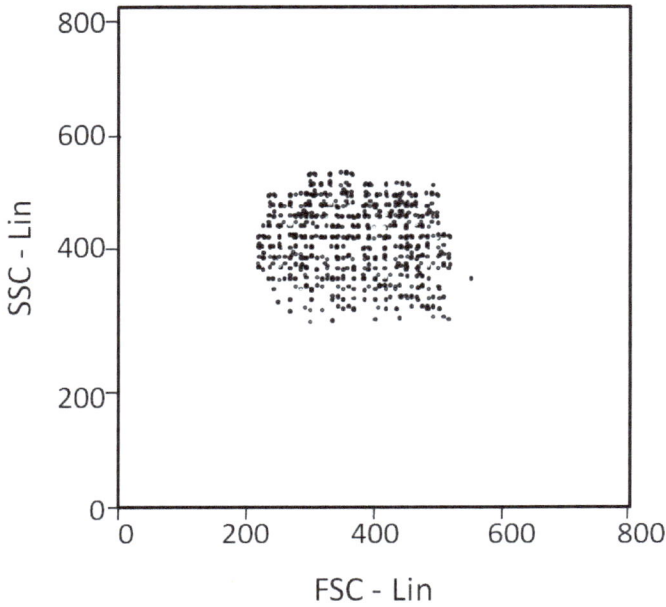

Fig. (6). Dot plot.

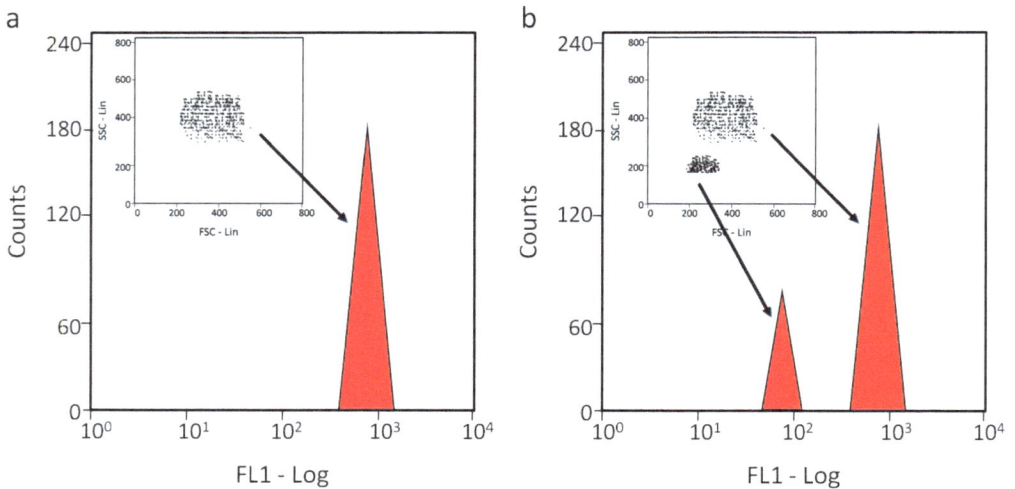

Fig. (7). (a) Histogram on the basis of a single type of population, **(b)** Histogram on the basis of two different types of populations.

Gates

A flow cytometer is unable to distinguish live cells from dead cells or debris. When the cell suspension is injected into a flow cytometer, these cells along with some debris or cellular clumps or dead cells encounter the laser, resulting in the visualization of the dot in dot plot data-sheet. This creates confusion whether all these are cells or not. These debris or dead cells can be distinguished on the basis of their size and scattering pattern (FSC and SSC). Usually, dead cells are smaller in size hence having lower FSC and high SSC as compared to live cells. This property is used to separate live cells from dead cells or debris in the dot plot data sheet. Such a mode of selection is called gating.

Illustration 2: In a flow cytometer experiment, white blood cells (WBCs) have been acquired. The overall picture on the screen has been depicted in Fig. (**8**). Granulocytes have more granules that are bigger in size, hence they have higher SSC and FSC as compared to other cells. Monocytes are larger in size but contain fewer granules hence they have higher FSC and lesser SSC. Being smaller in size with less granularity, lymphocytes are clustered well below the granulocytes and monocytes. All these cells can be gated on the basis of their characteristic FSC and SSC.

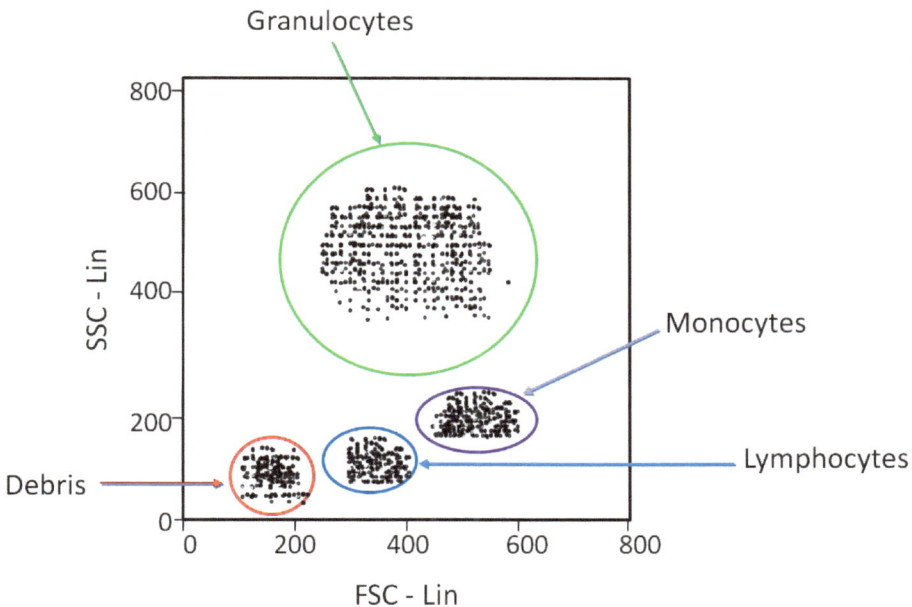

Fig. (8). Dot plot WBCs run into the flow cytometer. Each cell subset is scattered according to their FSC and SSC.

APPLICATIONS

Flow cytometer finds immense applications in biological sciences. A number of questions can be answered using the flow cytometer technique. Some of the applications of a flow cytometer in biomedical science are listed below.

1. Apoptosis Detection

Apoptosis is defined as a process of program cell death. Apoptosis is characterized by phosphatidylserine (PS) exposure over plasma membrane, change in mitochondrial permeability, DNA degradation and caspase activation [6]. The flow cytometer can be used to detect the apoptotic cells using any of the above-mentioned features. In apoptosis detection, the use of PS exposure is described. PS is a type of phospholipids distributed asymmetrically over the surface of the plasma membrane in live cells. During apoptosis, the PS flips towards the outer membrane by flippase enzyme-making cell which is symmetric in terms of PS exposure. Annexin V is a calcium-dependent molecule having a strong affinity with PS, hence it is used to detect the externalized PS. Annexin V is available as conjugated with FITC (Annexin V-FITC). Necrosis is defined as a form of accidental cell death which involves rupture of the plasma membrane. PI is a dye which is impermeable to early apoptotic cells whereas permeable to late apoptotic and necrotic cells. Hence a combination of Annexin V-FITC and PI is useful in the detection of apoptotic and necrotic cells in a population of cells.

Procedure: Apoptosis Detection Using Annexin-PI Labelling [7].

1. Wash 10^6 cells in PBS and resuspend in 1 mL of staining buffer.
2. Add 5 µl annexin V-FITC to the cell suspension and incubate for 10 minutes in the dark at 37°C.
3. Add 20 µl of PI to the cell suspension and further incubate for 10 minutes in dark.
4. Run the flow cytometer.
5. Viable cells are negative for both PI and annexin V, apoptotic cells are annexin V positive but PI negative and necrotic and late apoptotic cells are positive for annexin V and PI staining.

2. Cell Cycle Analysis

The mammalian cell cycle involves interphase and mitosis. The interphase is further sub-divided into G1, S, G2 and M phase. During mitosis, number of chromosomes or DNA content remains the same. However, in G2 phase, the DNA content is doubled compared to G1 phase. This is due to DNA replication occurring in the S phase. Deregulated cell cycle may lead to cancer or apoptosis. Therefore, analysis of the cell cycle is very important in biomedical science. There are various dyes (PI, Hoechst 33258, DRAQ5, and 7-amino actinomycin D) available that bind to the DNA and the fluorescence emitted is directly proportional to the DNA content. Hence, using these dyes in flow cytometer makes DNA content or cell cycle analysis easier [8]. PI is non-permeable to viable cells. Hence, cells need to be permeabilized prior to the use of PI. Since PI intercalates with the nucleic acids including DNA and RNA therefore, RNAse is used during sample preparation to get rid of RNA so that fluorescence signal get attributed to DNA only. Additionally, aneuploidy or polyploidy can also be detected using cell cycle analysis. Hence it is also useful in identifying chromosomal abnormalities.

Procedure: DNA Cell Cycle Analysis [9].

1. Harvest 10^5 cells by washing in PBS and pellet the cells by centrifugation.
2. Resuspend the pelleted cells in a hypotonic solution containing PI (50 µg/ml), RNase (50 µg/ml) and NP-40 (0.03%).
3. Incubate the reaction mixture at 4°C for 20 minutes in dark before running in the flow cytometer.
4. Set the fluorescence scale on linear mode during acquisition and analysis.
5. G2/M phase shows double fluorescence as compared to the G1 phase.
6. Fluorescence of the S phase lies in between G1 and G2/M phase (Fig. **9**).

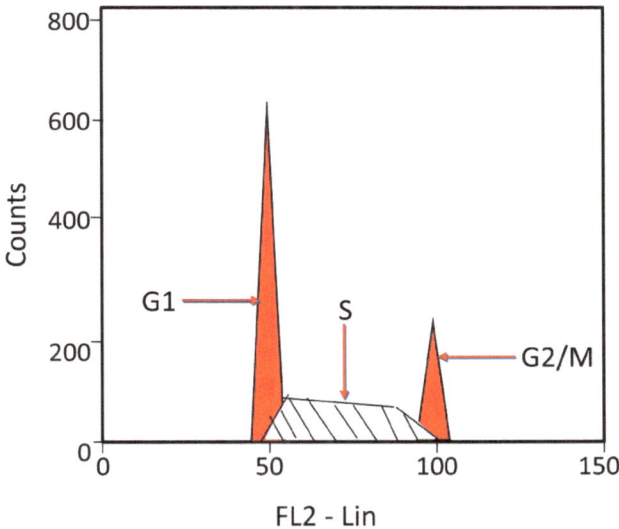

Fig. (9). Cell cycle stage of a eukaryotic cell.

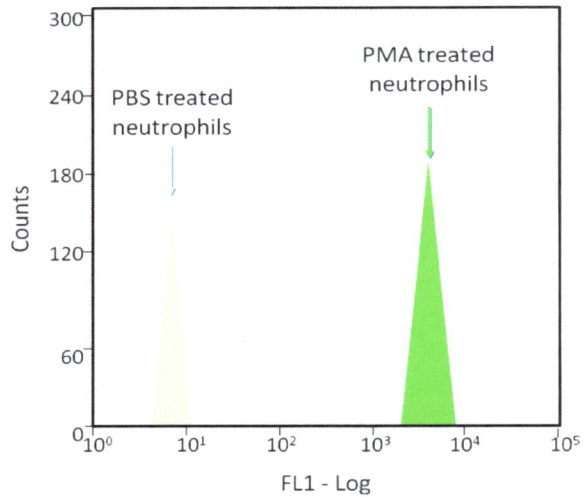

Fig. (10). Free radical generation in PBS and PMA treated neutrophils assessed by DCF-DA.

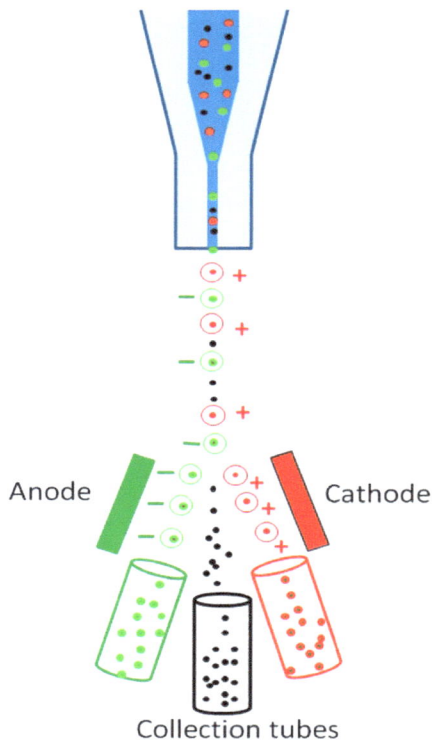

Fig. (11). .Sorting of CD4 and CD8 positive cells from WBCs by FACS.

3. Immunophenotyping

The identification of specific cells in a population of heterogeneous cells is an indispensable tool for disease diagnosis and classification of cells. Each cell has a unique marker in the form of proteins expressed over cell surface often called a cluster of differentiation (CD). The CD markers are highly specific to the corresponding cell types (CD15 for human neutrophils, CD4 for T helper cells, CD8 for T cytotoxic cells, CD25 for T regulatory cells, CD19 for B cells, CD14 for monocytes *etc.*). Flow cytometer-based detection of a specific cell is possible by using fluorochrome-conjugated specific antibodies against these CD markers [10]. Additionally, there are certain diseases that are characterized by an increase or decrease in the number of specific cell subset. In CD4 T helper cells the count is significantly decreased during HIV infection. Hence, the counting of CD4 T helper cells using anti-CD4 conjugated with fluorochrome in flow cytometer can be helpful in the diagnosis of AIDS.

Procedure: Identification of Human Neutrophils in Blood [11].

1. Harvest 10^6 blood cells in PBS.
2. Add 5µl of CD15-FITC antibody and a corresponding isotype control.
3. Incubate for 30 minutes at room temperature.
4. Wash cells by centrifugation and discard the supernatant.
5. Resuspend the cells in 0.5 ml PBS and place them in a flow cytometer.
6. Run the isotype control and set the baseline fluorescence by adjusting the voltage.
7. Run the CD15-FITC labeled cells in the same settings.
8. Calculate the percentage of cells showing CD15 positive using a software.

4. Free Radical Generation

Free radical species are defined as species having an unpaired electron. Several radical species including oxygen-centered radicals like superoxide anion, hydrogen peroxide, hydroxyl radical, and hypochlorous acid are produced during various signalling pathways involving microbicidal action by phagocytic cells (neutrophils and macrophages). These free radicals are produced by NADPH oxidase, myeloperoxidase, xanthine oxidase, and nitric oxide synthase. The measurement of free radicals is important in the detection of chronic granulomatous disease and myeloperoxidase deficiency. Flow cytometer-based detection of these radicals is possible using dyes that are specific to a particular radical. 2',7'-dichlorofluorescein diacetate (DCF-DA) is a nonfluorescent dye which enters inside the cell, and gets catalyzed by esterase enzyme and subsequently converts into DCF, a fluorescent probe. This DCF binds to the free radicals and emits green colour upon excitation with a blue laser [12]. If the fluorescence is high, the number of free radicals inside cells will be high as well.

Procedure: Detection of Free Radicals in Human Neutrophils [13].

1 Prepare neutrophil cells (10^6) and resuspend in PBS.
2 Add 5 µM of DCF-DA to the cells and incubate at 37°C in the dark for 20 minutes.
3 Add PBS (negative control) and 10 nM PMA (positive control) to separate tubes.
4 Incubate for 30 minutes at 37°C in the dark.
5 Wash the cells with PBS and place them in the flow cytometer.
6 Run the negative tube and set the auto-fluorescence at the minimal by adjusting the voltage.
7 Run the positive control tube on the same voltage and measure the fluorescence.
8 PMA treated neutrophils show enhanced fluorescence as compared to PBS treated cells (Fig. **10**).
9 Calculate the mean fluorescence using the software.

5. Cell Sorting

The sorting of cells from a heterogeneous population of cells is an important step for further study. Every cell has specific physical features characterized by their FSC and SSC in addition to the unique marker like CD present over the plasma membrane. Therefore, if a cell is identified using its FSC/SSC or CD marker, it can be retrieved using flow cytometer popularly known as fluorescence assisted cell sorter (FACS). Under ideal conditions, the sorted cells are nearly 99% pure. The major issue is low yield and there are chances of contamination [14]. To get more yields, a higher number of cells can be taken before sorting. Proper sterile conditions can be maintained to avoid contamination. The viability of cells can be maintained by adjusting the machine set up.

Procedure: Sorting of CD4+ and CD8+ Cells from WBCs [15].

1. Harvest the lymphocytes (10^8) and resuspend in PBS.
2. Add 5µl of CD4-PE and CD8-FITC antibody and a corresponding isotype control.
3. Incubate for 30 minutes at room temperature.
4. Wash cells by centrifugation and discard the supernatant.
5. Resuspend the cells in 5 ml RPMI media and place them in a flow cytometer.
6. Run the isotype control and set the baseline fluorescence by adjusting the

voltage.

7. Set up the machine so that each individual cell enters in the form of a single droplet.
8. Select the cells of interest on the basis of FITC and PE fluorescence
9. Give an electronic charge (say -ve charge for CD8-FITC cells and +ve charge for CD4-PE cells) to the cell inside the drop.
10. The positively charged droplets containing cell (CD4-PE) and negatively charged droplets containing cell (CD8-FITC) are deflected towards cathode and anode respectively.
11. Uncharged and unstained cells are collected separately in the centre tube.
12. The collection tubes below cathode will collect positively charged droplets having cells whereas, tube below anode will collect negatively charged droplets having cells (Fig. **11**).
13. Analyze the sorted cells and culture immediately to restore viability.

CONCLUDING REMARKS

Flow cytometry finds immense applications in the area of biomedical sciences from basic research to advanced clinical research. Flow cytometry is utilized in assessing apoptosis, free radical generation, cytokine estimation, DNA cell cycle analysis *etc*. In clinical laboratory, the flow cytometer is employed in detecting immunodeficiencies, lymphomas, leukaemias, and other immunophenotyping. The advanced techniques include cells sorting by FACS which enables sorting of a highly pure subset of cells from a mixture of a heterogeneous population. Being a sophisticated instrument, it requires training at a basic and advanced level so as to harness more diverse applications. This will not only increase the pace of basic cell biology research but it will also be helpful in disease diagnosis and hence increase patient care.

CONSENT FOR PUBLICATION

Not applicable.

CONFLICT OF INTEREST

The author confirm that the contents of this chapter have no conflict of interest.

ACKNOWLEDGEMENTS

Financial grant to Anupam Jyoti from the Department of Science and Technology, Science Engineering Research Board (YSS/2015/000846), New Delhi, India is highly acknowledged.

REFERENCES

[1]   Brunsting A, Mullaney PF. Differential light scattering from spherical mammalian cells. Biophys J 1974; 14(6): 439-53.
        [http://dx.doi.org/10.1016/S0006-3495(74)85925-4] [PMID: 4134589]

[2]   Crosland-Taylor PJ. A device for counting small particles suspended in a fluid through a tube. Nature 1953; 171(4340): 37-8.
        [http://dx.doi.org/10.1038/171037b0] [PMID: 13025472]

[3]   Mason WT, Ed. Fluorescent and luminescent probes for biological activity: a practical guide to technology for quantitative real-time analysis. Elsevier 1999.

[4]   Sebestyén Z, Nagy P, Horváth G, et al. Long wavelength fluorophores and cell-by-cell correction for autofluorescence significantly improves the accuracy of flow cytometric energy transfer measurements on a dual-laser benchtop flow cytometer. Cytometry 2002; 48(3): 124-35.
        [http://dx.doi.org/10.1002/cyto.10121] [PMID: 12116358]

[5]   Nguyen DC, Keller RA, Jett JH, Martin JC. Detection of single molecules of phycoerythrin in hydrodynamically focused flows by laser-induced fluorescence. Anal Chem 1987; 59(17): 2158-61.
        [http://dx.doi.org/10.1021/ac00144a032] [PMID: 19886672]

[6]   Allen PD, Bustin SA, Newland AC. The role of apoptosis (programmed cell death) in haemopoiesis and the immune system. Blood Rev 1993; 7(1): 63-73.
        [http://dx.doi.org/10.1016/0268-960X(93)90025-Y] [PMID: 8467234]

[7]   Kumar S, Barthwal MK, Dikshit M. Cdk2 nitrosylation and loss of mitochondrial potential mediate NO-dependent biphasic effect on HL-60 cell cycle. Free Radic Biol Med 2010; 48(6): 851-61.
        [http://dx.doi.org/10.1016/j.freeradbiomed.2010.01.004] [PMID: 20079829]

[8]   Merkel DE, Dressler LG, McGuire WL. Flow cytometry, cellular DNA content, and prognosis in human malignancy. J Clin Oncol 1987; 5(10): 1690-703.
        [http://dx.doi.org/10.1200/JCO.1987.5.10.1690] [PMID: 3309200]

[9]   Kumar S, Barthwal MK, Dikshit M. Nitrite-mediated modulation of HL-60 cell cycle and proliferation: involvement of cyclin-dependent kinase 2 activation. J Pharmacol Exp Ther 2011; 337(3): 312-21.
        [http://dx.doi.org/10.1124/jpet.110.177444] [PMID: 21411497]

[10]   Passlick B, Flieger D, Ziegler-Heitbrock HW. Identification and characterization of a novel monocyte subpopulation in human peripheral blood. Blood 1989; 74(7): 2527-34.
        [PMID: 2478233]

[11]   Fornas O, Garcia J, Petriz J. Flow cytometry counting of CD34+ cells in whole blood. Nat Med 2000; 6(7): 833-6.
        [http://dx.doi.org/10.1038/77571] [PMID: 10888936]

[12]   Dikshit M, Sharma P. Nitric oxide mediated modulation of free radical generation response in the rat polymorphonuclear leukocytes: a flowcytometric study In Advanced Flow Cytometry: Applications in Biological Research. Dordrecht: Springer 2003; pp. 69-76.

[13]   Patel S, Kumar S, Jyoti A, et al. Nitric oxide donors release extracellular traps from human neutrophils by augmenting free radical generation. Nitric Oxide 2010; 22(3): 226-34.
        [http://dx.doi.org/10.1016/j.niox.2010.01.001] [PMID: 20060922]

[14]   Cormack BP, Valdivia RH, Falkow S. FACS-optimized mutants of the green fluorescent protein (GFP). Gene 1996; 173(1 Spec No): 33-8.
        [http://dx.doi.org/10.1016/0378-1119(95)00685-0] [PMID: 8707053]

[15]   Magbanua MJ, Park JW. Isolation of circulating tumor cells by immunomagnetic enrichment and fluorescence-activated cell sorting (IE/FACS) for molecular profiling. Methods 2013; 64(2): 114-8.
        [http://dx.doi.org/10.1016/j.ymeth.2013.07.029] [PMID: 23896286]

CHAPTER 4

Protein Structure Determination Using Experimental Phasing Technique

Vijay Kumar Srivastava[*]

Amity Institute of Biotechnology, Amity University Rajasthan, Kant Kalwar, NH-11C, Jaipur-Delhi Highway, Jaipur-303002, India

Abstract: In X-ray crystallography, the phase problem is a major bottle neck, it would be successful only if we obtain a suitable template or phases can be determined experimentally. However molecular replacement calculation is the most useful technique and it could be applied when there is a homologous structure known. To obtain the phase information, the most widely used method is derivatization of heavy-atoms for protein crystals, although it is generally not used nowadays. While it is a valuable technique to obtain the phases for unidentified structures having no identity with the homologous for using molecular replacement (MR) and the crystals that cannot be diffracted even at synchrotron, nowadays, the SAD and MAD methods for experimental phasing have been developed. The more advanced technique has also been developed that is Cryo-EM and SFX with XFELs, having enormous potential for determining the structure of novel proteins that are not acquiescent to produce a crystal that cannot diffract.

Keywords: Cryo-electron microscopy (Cryo-EM), Experimental phasing, Multiwavelength anomalous dispersion, MAD (multiple-wavelength anomalous dispersion), Single wavelength anomalous dispersion, Single and multiple isomorphous replacement, Serial femtosecond crystallography (SFX) with X-ray free electron lasers (XFELs), SAD (single-wavelength anomalous dispersion).

INTRODUCTION

Experimental phasing technique is one of the most difficult and time taking processes in comparison to that of solving the structure using the molecular replacement calculation. The former technique is essential when an unknown structure cannot be solved and has a novel fold. Since these unknown macromolecular structures deal with the new function that correlates the biology,

[*] **Corresponding author: Vijay Kumar Srivastava:** Amity Institute of Biotechnology, Amity University Rajasthan, Kant Kalwar, NH-11C, Jaipur-Delhi Highway, Jaipur-303002, India; E-mails: vksrivastava@jpr.amity.edu and vijaytechno@gmail.com

Anupam Jyoti & Neetu Mishra (Eds.)

experimental phasing remains a useful technique in the field of X-ray crystallography [1]. This book chapter includes a basic understanding of the experimental phasing and how it will be used in the simplest way. It also includes data reduction, model building, and refinement during the experimental phasing technique. In addition to that, the book chapter also deals with the advanced techniques used in macromolecular X-ray crystallography.

EXPERIMENTAL PHASING METHODS USING HEAVY ATOMS (HAS)

Macromolecules have no or low sequence identity with known homologous structures for molecular replacement calculations and the crystals that do not diffract; where initial phases have to be obtained using the experimental phasing techniques. This will be achieved by soaking the heavy atoms with the crystals, if soaked heavy atoms are placed in a periodic manner in each asymmetric unit in the crystal. Heavy atoms always have a higher number of electrons than those ordinarily found in macromolecules (C, H, N, O and S atoms). Diffraction of X-rays relies upon the number of electrons present outside the nucleus in an atom, so the x-ray diffraction will be greater for the heavier atoms. The phase is determined by obtaining the difference between the native and heavy atoms soaked crystals [2]. Recently, a new method has been developed by a crystallographer to solve structures of unknown molecules by exploring the anomalous signal from the wild type sulfur (S) atoms present in methionine and cysteine residues [3 - 6], although S atoms are not considered as a heavy atom. In conclusion, however, it is likely that a heavy atom phasing is required to solve the structure of a macromolecule having no structural homologous.

REPLACEMENT OF SULFUR ATOMS WITH SELENIUM

The substitution of S atoms with selenium is the most advanced and widely accepted technique of experimental phasing [3, 7, 8]. This technique works in those proteins that are easily expressed and purified in prokaryotic system for example *E. coli* [9 - 11]. In *E. coli* expression systems, the methionine auxotroph strain B834 (DE3) of *E. coli* or normal *E. coli* strain is grown on minimal medium with selenomethionine as the exclusive source of methionine. The protein expressed in a particular system has an increased number of its methionines replaced by selenomethionines and further it can be proved by mass spectrometry analysis [12]. The insect cells can also be used to produce selenomethionine-labeled protein, despite methionine-auxotroph insect-cell strains are not present, so substitution depends on the replacement of the medium with minimal medium consisting of selenomethionine, leading to an inadequate exchange [13].The expression level, crystallization, stability, and selenomethionine (SeMet) labeled proteins diffraction vary from species to species and it is not always sure whether these will work in all the cases. Selenium atom has low X-ray scattering as compared to the heavy atoms for *e.g.* platinum (78 electrons), so to obtain the phases accurately with SeMet, the protein should have more methionines for substitution [14].

STRUCTURE SOLUTION

Crystals are the repeating units of ordered atoms with this minimum energy state in three directional direct/real space (xyz). The repeating units are known as unit cells. The real space is represented by the electron density, $\rho(xyz)$, a function defined in three-dimensional xyz coordinates. X-rays collide with the atomic electrons in the crystals resulting in a diffraction pattern, also called reciprocal space [15]. The direct and reciprocal spaces are related by a mathematical term, called Fourier transform. The diffraction experiments provide the intensities of the waves scattered through the hkl planes that represent the miller indices of the diffracted beam in the crystal which reflects the number of electrons present in one particular plane. The amplitude of the scattered wave, F(hkl), also known as the structure factor, is directly proportional to the square root of the diffracted intensities. The electron density at the xyz position in the unit cell is defined by the additive phenomena of a scattered wave from hkl plane to the point xyz, whose amplitude depends on the number of the electron in the place as well as the relative phases. Hence, to calculate the electron density distribution or alternatively to detect the atomic positions inside the unit cell we need the F(hkl) and the phase information of the diffracted beam α(hkl) as described below:

$$\rho(x, y, z) = 1/V \; \Sigma_{hkl} \; F(h \; k \; l) \; \exp \; [-2\pi i(hx + ky + lz) + i\alpha(h \; k \; l)]$$

Where, V represents the volume of the unit cell, F(h k l) is the structure factor for the equivalent reflection and α(h k l) are the phases for each point *xyz*, in the three-dimensional unit cell [15].

With the diffraction experiment, we only obtain information of measured intensity. The phase information of the diffracted beam α_{hkl} is lost during the diffraction experiment. Although there is no direct relationship between the amplitudes and phases, the initial phase information can be deduced using several techniques; Molecular replacement (MR), Isomorphous replacement (SIR and MIR) and Anomalous dispersion (SAD and MAD).

Multiwavelength anomalous dispersion(MAD) information from anomalously scattering atom forms the basis of structure determination.

Multiple isomorphous replacement method (MIR) requires the incorporation of heavy atoms in protein crystals. The native and heavy atoms incubated crystals can be diffracted and the phase information can be obtained by the difference observed in the diffraction intensities. In order to do so, the position and occupancy of the heavy atoms are identified using difference Patterson function. Multiple isomorphous replacements, which involve the following steps, are performed by respective programs:

Determining Heavy Atom Positions

In this technique, the contribution of heavy atoms in phase and amplitude to the structure factor of the heavy atoms derivative has to be identified. The position of change can be determined first from the observed difference in scattering amplitudes caused by incorporation of the heavy atoms. The different techniques to identify the position of the heavy atom are as follows:

• Patterson searches
• Direct methods
• Difference Fourier method.

All the above techniques require a previous calculation of F_H obtained from observed differences, this first one involves the scaling of native and derivative data sets and magnitudes for amplitudes of heavy atoms contribution can be obtained from the experimental data but phase of F_H is obtained from either centric or acentric reflections, where F_H is related to F_P and F_{PH} by vector equation:

$$F_H = F_{PH} - F_P$$

Where F_P and F_{PH} are structure factors of native and derivative, respectively.

I. **Patterson Search**: Patterson's method of search is more general method and does not require any previous Pattersons.

II. **Direct Methods**: In this method, the position of heavy atoms is difficult to identify by the difference Patterson, so it is convenient to use:

  I. SHELX [16].
 II. MULTAN: uses the observed anomalous scattering [17].
III. **Difference Fouriers:** It requires the information of the previous phase of protein.

Refinements of Heavy Atom Parameters

A heavy atom has to be refined in several steps before it is used for phase determination.

(i) **Maximum Likelihood phase (ML):** In this method, we refine the coordinates, occupancy and thermal factor of the heavy atoms [18].

Phase Determination

(i) **Isomorphous Replacement:** From the vector triangle F_{PH} = Fp + F_H, of the amplitudes F_{PH}, Fp and F_H and the phase α_P can be derived by applying the cosine law; α_P can be derived as follows:

$$\alpha_P = \alpha_H + \cos^{-1}((F_{PH}^{\ 2} - F_P^{\ 2} - F^2_{\ H})/2F_PF_H)$$
$$= \alpha_H \pm \varphi$$

To identify a clear estimate of α_P, it is needed to repeat the phase determination using another vector FH_2.

(ii) **Anomalous Scattering:** To identify the phase using anomalous scattering, it is truly based on the concept that there will be a change in the amplitude and phase when the heavy atom is able to scatter X-rays at anomalously wavelength. The resultant change that occurs in the diffracted intensities is applied to find the heavy atom substructure. Furthermore, the phases are estimated for all the structure factor amplitudes.

(iii) **Density Modification:** It is used to improve the isomorphous phases in which the density map is modified accordingly with the information of protein structure to obtain a more improved map [19].

Software Used for Structure Solution through MIR

SOLVE/RESOLVE: This technique is used for structural determination and it involves all the steps of structure solution that is data reduction, locating the position of the heavy atom followed by modification of electron density and model building. SOLVE is used to locate the position of heavy atoms in macromolecular structure solution. RESOLVE identifies the non crystallography symmetry (NCS), density modification to improve the phases and automated model building. The first step in the SOLVE is the scaling of data sets by merging all the measurements into an asymmetric unit of the crystal. The second step is the generation of solution for anomalous difference Patterson function. Third, it identifies the position of the heavy atom and ranks them. RESOLVE in SOLVE/RESOLVE basically performs statistical modification of density and automated model-building [20 - 22].

Structure Refinement and Validation

After obtaining the phases, the model is built and the refinements are made by optimizing the xyz coordinates to try to fit the molecule in an electron density map. Model building and refinement are repeated several times until and unless there is an agreement between the calculated and the observed structure factors. The assessment of the refined final model is made by several factors which are as follows:

- Goodness of the electron density maps.
- R_{factor} and R_{free}
- Ramachandran plot

$$R_{factor} = \Sigma hkl \mid \mid F_o \mid - k \mid F_c \mid \mid / \Sigma hkl \mid F_o \mid$$

Where F_o is observed and F_c is calculated structure factors.

Different Programs to Refine the Macromolecular Structures, are as follows:

REFMAC5: It is a technique that is based on maximum likelihood refinement [23].

Automated Refinement Procedure (**Arp/warp**): It identifies of electron density map and performs automated model building and refinement [24].

Xplor: X-ray and molecular dynamics refinement program. This can be used to perform simulated annealing, conjugate gradient minimization *etc* [25]. **Phenix. refine:** is an advanced program that includes all the parameters for refinement of the protein structure [23]. The experimental data given as input are checked for outliers [26].

REFMAC5, implemented in the CCP4 program suit and phenix, is mainly used for the refinement nowadays, is described below:

REFMAC5

It is the process of manual model building and refinement to obtain a good concurrence between the calculated and the observed structure factors. The refinement mechanism implemented in REFMAC5 consists of three steps:

- Initialization- the processing of input data and parameters and crystallographic information files for small molecules.
- Repeated refinement of the bulk solvent and scaling, xyz coordinates temperature factor and simulated annealing *etc.*
- Output-comprehensive summary of the refinement steps where the refined model, electron density maps and many statistics are reported.

Structure refinement consists of several steps, in which each step specifically involves iteration of coordinate refinement, real and rigid refinement, simulated annealing and torsion angle dynamics followed by grouped or individual B-factor refinement. The standard parameters used in the current study are:

- **R-factor**- indicates the accuracy of the model. The use of free R-factor during the refinement is a sign of model quality and thus a statistical cross-validation approach.
- **Maximum likelihood refinement target function (ML)**- It gives the better target for macromolecular refinement by computing incompleteness of the model that is missing atoms.
- **Rigid body minimization**- It refines the three rotational and three translational degrees of freedoms of the specified rigid groups by minimizing the contrast between the observed and calculated structure factors.
- **Refinement of XYZ coordinate** - It involves the cartesian refinement of model geometry. Often the additional restrains, like NCS and reference model, are used along with the coordinate refinement to prevent the over-fitting.
- **Real-space refinement**- Using the electron density maps as a target, it refines the atomic positions. This is useful for both local (fitting individual residues and rotamer outliers) as well as global refinement.
- **Grouped B-factor refinement**- This parameter is employed for the crystallographic data at low resolution which restricts the data to parameter ratio. It refines the B-factors for multiple atoms of a residue at a time including both the main chain and side chain by minimizing the E_{xray} function.
- **Simulated annealing**- Annealing is the process in which the molecules are heated to its highest temperature and then the system is slowly cooled down so that they come to the lowest energy state [25, 27]. The major advantage of this procedure is that it provides a significant improvement in the model refinement by removing phase biases and overcoming potential barriers.
- **NCS restraints**- It is used whenever the molecule has non crystallographic symmetry present in the protein crystal or unit cell [23].
- **Secondary structure restraints**- It refines the hydrogen bonds in alpha helices and beta sheets required for maintaining correct geometry at a lower resolution [19].
- **Constraints and restraints**- During the refinement, constraints and restraints are applied on the bond lengths, atomic coordinates, bond angles and NCS by using force constants to improve the model. Given limited freedom on a parameter, restraints are applied, while for an exact parameter value, constraints are used. Thus, a constraint is restraint with an infinite force constant.
- **Bulk solvent scattering**- The bulk solvent correction during the refinement is used to compensate for scattering at low resolution. The volume filled by solvent

is identified by demarcating a solvent-accessible volume outside the Van der Waal's exclusion zone of the protein.

Phenix Refinement

It involves the following steps:

- The input data are stored in PDB format and cif files with stereochemistry definitions.
- The bulk solvent correction, refinement of XYZ coordinates, temperature-factors and water picking.
- Rigid body refinement [28].
- Simulated-annealing refinement in cartesian or torsion angle space [29].
- Detection of non crystallographic symmetry and its refinement, especially used for low-resolution structures.
- Identification of Ramachandran Plot using the MolProbity feature [30].

Fourier Maps and Map Interpretation

After every round of refinement, the model is checked and manually missing residues are fitted into the electron density map using the program COOT [31]. The free R- factor [25] is checked at each stage to prevent model biases. In the final round of refinement, solvent molecules are added automatically by COOT [31]. At the final stage, the model is checked manually to identify the excess water molecule on the basis of electron density maps.

Omit Maps

Omit maps are used to remove systematic errors arising out of model bias due to the initial model used in the structure solution, and/or subsequent refinement. This is done by refining the molecule after omitting the region of interest. A subsequently calculated map is largely free from the model bias in the omitted region. One of the ways to do this is through the use of 'Omit-type' maps [32].

Geometric Analysis

REFMAC program analyses the deviations in the bond lengths and bond angles, short contacts between the symmetry-related atoms, *etc.* The program also lists the energies that deviate from weights used for the refinement [33]. Average B-factors for the protein atoms, ligands, cofactor, and water molecule were determined by using the BAVERAGE program. Interactions are calculated using the program CONTACT module of CCP4 [34].

PROCHECK: This program [35], analyses the stereochemical quality of the protein structures. It forms a part of the CCP4 suite of programs [34]. The output consists of a number of 'postscript' files and comprehensive residue-by-residue listing of the parameters. It highlights the regions of the model that may need further investigation. It compares and assesses the quality of the model with regard to the other structures at comparable resolutions. It also suggests corrections in the atom numbering in line with the guidelines given by the IUPAC-IUB Commission on Biochemical Nomenclature, 1970. PROCHECK was used to determine the quality of the model after every round of refinement.

Superposition of Structures: All the x-ray structural superimpositions and *r.m.s.d.* for macromolecular structures are determined using the program ALIGN (https://zhanglab.ccmb.med.umich.edu/MM-align) and PROFIT (Martin, A. C. R).The whole methodology for the chapter is shown in Fig. (**1**)

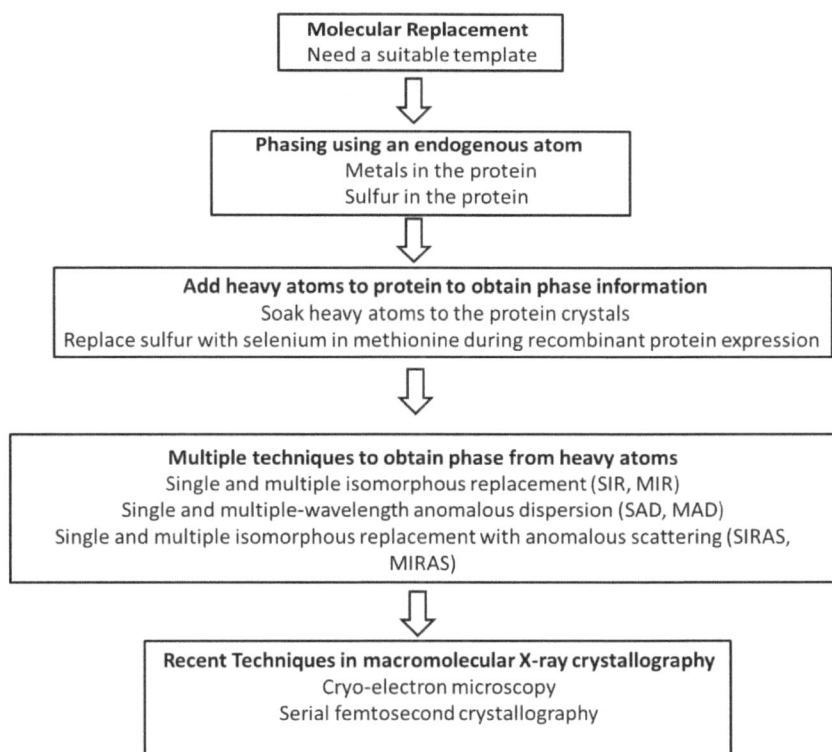

```
┌─────────────────────────────────────┐
│        Molecular Replacement         │
│        Need a suitable template      │
└─────────────────────────────────────┘
                  ⇩
┌─────────────────────────────────────┐
│   Phasing using an endogenous atom   │
│        Metals in the protein         │
│        Sulfur in the protein         │
└─────────────────────────────────────┘
                  ⇩
┌───────────────────────────────────────────────────────────┐
│   Add heavy atoms to protein to obtain phase information    │
│        Soak heavy atoms to the protein crystals             │
│ Replace sulfur with selenium in methionine during recombinant protein expression │
└───────────────────────────────────────────────────────────┘
                  ⇩
┌───────────────────────────────────────────────────────────┐
│     Multiple techniques to obtain phase from heavy atoms    │
│   Single and multiple isomorphous replacement (SIR, MIR)    │
│ Single and multiple-wavelength anomalous dispersion (SAD, MAD) │
│ Single and multiple isomorphous replacement with anomalous scattering (SIRAS, MIRAS) │
└───────────────────────────────────────────────────────────┘
                  ⇩
┌───────────────────────────────────────────────────────────┐
│ Recent Techniques in macromolecular X-ray crystallography   │
│           Cryo-electron microscopy                          │
│        Serial femtosecond crystallography                   │
└───────────────────────────────────────────────────────────┘
```

Fig. (1). Different methods used for the phasing of X-ray structures.

Current Advancements in Macromolecular X-ray Crystallography

Cryogenic electron microscopy (cryo-EM) is generally known for identification of vitreous ice-embedded macromolecular complexes in a cryogenic transmission electron microscopy (TEM). It aims to identify the three-dimensional structures of the macromolecular complexes that are truly based on their extremely noisy two-dimensional projection images. The liquid helium is used to freeze the biological sample given by Fernandez-Moran [36]. Dubochet and his colleagues discovered the method for "single particles" known as specimen vitrification by quickly plunging the sample into liquid ethane that saves it from structural damage and by electron beam. Originally, the computational methods, simultaneously derived from the three-dimensional X-ray crystallography, could only deal with the highest regular specimens, such as 2-dimensional crystals, helical fibers, and icosahedral assemblies. This overall method is known as single-particle cryo-EM when it is used for frozen-hydrated specimens. Several further developments in EM hardware are still necessary for the advancement of this method.

SFX with XFELs is an advanced technique for identifying the structure of novel and membrane proteins that are not able to produce crystals, that can be diffracted at a higher resolution. All in all, novel techniques and hardware form the basis of the present "resolution revolution" we noticed in the field of protein X-ray crystallography.

CONCLUDING REMARKS

The major bottle neck in the area of macromolecular X-ray crystallography is determining the phases. To get the phases for a protein having low sequence identity or no homologous is a difficult task. In order to get the phases for the protein having novel or newer fold, the experimental phasing techniques are used. Nowadays, the most common method of experimental phasing technique used is selenomethionine (SeMet) phasing or sulfur atom phasing. One SeMet is required to obtain a phase for 100 residues protein. These could then be used to optimize diffraction quality and subsequently be used for SeMet labeling. However, in many cases, crystals do not diffract and SeMet method fails, when they are used for large macromolecular complexes or membrane proteins. To overcome the problem of getting diffracting protein crystals, nowadays cryo-EM and SFX with X-ray free electron lasers (XFELs) are widely used. Still there is a need to optimize the process of cryo-EM that is grid preparation to TEM assessment and data collection, trial and-error process that needs considerable expertise, time, and consumable supplies. The advanced generation of TEM instruments can load a dozen cryo-EM grids in parallely, which overcomes the problem of handling the grid manually. However in near future, we can assume that cryo-EM would be fully automated from the preparation of the grid and data collection, so that it is used as high-throughput method.

CONSENT FOR PUBLICATION

Not applicable.

CONFLICT OF INTEREST

The author confirms that these chapter contents have no conflict of interest.

ACKNOWLEDGMENTS

We extend our appreciation towards Amity University Rajasthan, Jaipur, India for its valuable support throughout the work.

REFERENCES

[1]   Garman EF. Developments in x-ray crystallographic structure determination of biological macromolecules. Science 2014; 343(6175): 1102-8.
[http://dx.doi.org/10.1126/science.1247829] [PMID: 24604194]

[2]   Boggon TJ, Shapiro L. Screening for phasing atoms in protein crystallography. Structure 2000; 8(7): R143-9.
[http://dx.doi.org/10.1016/S0969-2126(00)00168-4] [PMID: 10903954]

[3]   Hendrickson WA, Horton JR, LeMaster DM. Selenomethionyl proteins produced for analysis by multiwavelength anomalous diffraction (MAD): a vehicle for direct determination of three-dimensional structure. EMBO J 1990; 9(5): 1665-72.
[http://dx.doi.org/10.1002/j.1460-2075.1990.tb08287.x] [PMID: 2184035]

[4]   Debreczeni JE, Bunkóczi G, Ma Q, Blaser H, Sheldrick GM. In-house measurement of the sulfur anomalous signal and its use for phasing. Acta Crystallogr D Biol Crystallogr 2003; 59(Pt 4): 688-96.
[http://dx.doi.org/10.1107/S0907444903002646] [PMID: 12657788]

[5]   Sarma GN, Karplus PA. In-house sulfur SAD phasing: a case study of the effects of data quality and resolution cutoffs. Acta Crystallogr D Biol Crystallogr 2006; 62(Pt 7): 707-16.
[http://dx.doi.org/10.1107/S0907444906014946] [PMID: 16790926]

[6]   Goulet A, Vestergaard G, Felisberto-Rodrigues C, *et al.* Getting the best out of long-wavelength X-rays: de novo chlorine/sulfur SAD phasing of a structural protein from ATV. Acta Crystallogr D Biol Crystallogr 2010; 66(Pt 3): 304-8.
[http://dx.doi.org/10.1107/S0907444909051798] [PMID: 20179342]

[7]   Walden H. Selenium incorporation using recombinant techniques. Acta Crystallogr D Biol Crystallogr 2010; 66(Pt 4): 352-7.
[http://dx.doi.org/10.1107/S0907444909038207] [PMID: 20382987]

[8]   Metanis N, Hilvert D. Natural and synthetic selenoproteins. Curr Opin Chem Biol 2014; 22: 27-34.
[http://dx.doi.org/10.1016/j.cbpa.2014.09.010] [PMID: 25261915]

[9]   Cronin CN, Lim KB, Rogers J. Production of selenomethionyl-derivatized proteins in baculovirus-infected insect cells. Protein Sci 2007; 16(9): 2023-9.
[http://dx.doi.org/10.1110/ps.072931407] [PMID: 17660253]

[10] Nettleship JE, Assenberg R, Diprose JM, Rahman-Huq N, Owens RJ. Recent advances in the production of proteins in insect and mammalian cells for structural biology. J Struct Biol 2010; 172(1): 55-65.
[http://dx.doi.org/10.1016/j.jsb.2010.02.006] [PMID: 20153433]

[11] Srivastava VK, Chandra M, Saito-Nakano Y, Nozaki T, Datta S. Crystal Structure Analysis of Wild

Type and Fast Hydrolyzing Mutant of EhRabX3, a Tandem Ras Superfamily GTPase from Entamoeba histolyica. J Mol Biol 2016; 428(1): 41-51.
[http://dx.doi.org/10.1016/j.jmb.2015.11.003] [PMID: 26555751]

[12]   Garman EF, Grime GW. Elemental analysis of proteins by microPIXE. Prog Biophys Mol Biol 2005; 89(2): 173-205.
[http://dx.doi.org/10.1016/j.pbiomolbio.2004.09.005] [PMID: 15910917]

[13]   Bellizzi JJ, Widom J, Kemp CW, Clardy J. Producing selenomethionine-labeled proteins with a baculovirus expression vector system. Structure 1999; 7(11): R263-7.
[http://dx.doi.org/10.1016/S0969-2126(00)80020-9] [PMID: 10574801]

[14]   Kumar Srivastava V, Chandra M, Datta S. Crystallization and preliminary X-ray analysis of RabX3, a tandem GTPase from Entamoeba histolytica. Acta Crystallogr F Struct Biol Commun 2014; 70(Pt 7): 933-7.
[http://dx.doi.org/10.1107/S2053230X14011388] [PMID: 25005092]

[15]   Scapin G. Molecular replacement then and now. Acta Crystallogr D Biol Crystallogr 2013; 69(Pt 11): 2266-75.
[http://cx.doi.org/10.1107/S0907444913011426] [PMID: 24189239]

[16]   Thorn A. Experimental Phasing: Substructure Solution and Density Modification as Implemented in SHELX. Methods Mol Biol 2017; 1607: 357-76.
[http://dx.doi.org/10.1007/978-1-4939-7000-1_15] [PMID: 28573581]

[17]   Ramagopal UA, Dauter M, Dauter Z. SAD manganese in two crystal forms of glucose isomerase. Acta Crystallogr D Biol Crystallogr 2003; 59(Pt 5): 868-75.
[http://dx.doi.org/10.1107/S0907444903005663] [PMID: 12777803]

[18]   Otwinowski Z, Minor W. Processing of X-ray diffraction data collected in oscillation mode. Methods Enzymol 1997; 276: 307-26.
[http://dx.doi.org/10.1016/S0076-6879(97)76066-X]

[19]   Cowtan K. Recent developments in classical density modification. Acta Crystallogr D Biol Crystallogr 2010; 66(Pt 4): 470-8.
[http://dx.doi.org/10.1107/S090744490903947X] [PMID: 20383000]

[20]   Terwilliger TC, Berendzen J. Automated MAD and MIR structure solution. Acta Crystallogr D Biol Crystallogr 1999; 55(Pt 4): 849-61.
[http://dx.doi.org/10.1107/S0907444999000839] [PMID: 10089316]

[21]   Terwilliger TC. Maximum-likelihood density modification. Acta Crystallogr D Biol Crystallogr 2000; 56(Pt 8): 965-72.
[http://dx.doi.org/10.1107/S0907444900005072] [PMID: 10944333]

[22]   Terwilliger TC. Automated main-chain model building by template matching and iterative fragment extension. Acta Crystallogr D Biol Crystallogr 2003; 59(Pt 1): 38-44.
[http://dx.doi.org/10.1107/S0907444902018036] [PMID: 12499537]

[23]   Murshudov GN, Vagin AA, Lebedev A, Wilson KS, Dodson EJ. Efficient anisotropic refinement of macromolecular structures using FFT. Acta Crystallogr D Biol Crystallogr 1999; 55(Pt 1): 247-55.
[http://dx.doi.org/10.1107/S090744499801405X] [PMID: 10089417]

[24]   Perrakis A, Morris R, Lamzin VS. Automated protein model building combined with iterative structure refinement. Nat Struct Biol 1999; 6(5): 458-63.
[http://dx.doi.org/10.1038/8263] [PMID: 10331874]

[25]   Brünger AT. Crystallographic refinement by simulated annealing. Application to a 2.8 A resolution structure of aspartate aminotransferase. J Mol Biol 1988; 203(3): 803-16.
[http://dx.doi.org/10.1016/0022-2836(88)90211-2] [PMID: 3062181]

[26]   Read RJ. Pushing the boundaries of molecular replacement with maximum likelihood. Acta Crystallogr D Biol Crystallogr 2001; 57(Pt 10): 1373-82.

[http://dx.doi.org/10.1107/S0907444901012471] [PMID: 11567148]

[27]   Brünger AT, Adams PD, Clore GM, *et al.* Crystallography & NMR system: A new software suite for macromolecular structure determination. Acta Crystallogr D Biol Crystallogr 1998; 54(Pt 5): 905-21. [http://dx.doi.org/10.1107/S0907444998003254] [PMID: 9757107]

[28]   Afonine PV, Grosse-Kunstleve RW, Adams PD. A robust bulk-solvent correction and anisotropic scaling procedure. Acta Crystallogr D Biol Crystallogr 2005; 61(Pt 7): 850-5. [http://dx.doi.org/10.1107/S0907444905007894] [PMID: 15983406]

[29]   Brunger AT, Kuriyan J, Karplus M. Crystallographic R factor refinement by molecular dynamics. Science 1987 Jan 23; 235(4787): 458-60.

[30]   Headd JJ, Immormino RM, Keedy DA, Emsley P, Richardson DC, Richardson JS. Autofix for backward-fit sidechains: using MolProbity and real-space refinement to put misfits in their place. J Struct Funct Genomics 2009; 10(1): 83-93. [http://dx.doi.org/10.1007/s10969-008-9045-8] [PMID: 19002604]

[31]   Emsley P, Cowtan K. Coot: model-building tools for molecular graphics. Acta Crystallogr 2004 Dec; 60(Pt 12 Pt 1): 2126-32. [http://dx.doi.org/10.1107/S0907444904019158]

[32]   Terwilliger TC, Grosse-Kunstleve RW, Afonine PV, *et al.* Iterative model building, structure refinement and density modification with the PHENIX AutoBuild wizard. Acta Crystallogr D Biol Crystallogr 2008; 64(Pt 1): 61-9. [http://dx.doi.org/10.1107/S090744490705024X] [PMID: 18094468]

[33]   Leonard AN, Pastor RW, Klauda JB. Parameterization of the CHARMM All-Atom Force Field for Ether Lipids and Model Linear Ethers. J Phys Chem A (Jun): 21.

[34]   Yu-dong L, Harvey I, Yuan-xin G, *et al.* Is single-wavelength anomalous scattering sufficient for solving phases? A comparison of different methods for a 2.1 A structure solution. Acta Crystallogr D Biol Crystallogr 1999; 55(Pt 9): 1620-2. [http://dx.doi.org/10.1107/S0907444999007726] [PMID: 10489467]

[35]   Laskowski RA. PDBsum: summaries and analyses of PDB structures. Nucleic Acids Res 2001; 29(1): 221-2. [http://dx.doi.org/10.1093/nar/29.1.221] [PMID: 11125097]

[36]   Fernandez-Moran H. Low-temperature preparation techniques for electron microscopy of biological specimens based on rapid freezing with liquid helium II. Ann N Y Acad Sci 1960; 85: 689-713. [http://dx.doi.org/10.1111/j.1749-6632.1960.tb49990.x] [PMID: 13698977]

Essentials of Recombinant Protein Production

Satyajeet Das and **Sanket Kaushik**[*]

Amity Institute of Biotechnology, Amity University Rajasthan, Rajasthan, India

Abstract: With the increasing demand for commercially important proteins, it has become essential to develop more improved and efficient methods for heterologous protein production to meet the requirements of society. Although currently there are different host systems available for the production of proteins heterologously, still the selection of a perfect host system becomes a tedious task if the aim is to get fully functional protein in adequate quantity. Different types of hosts are commonly used for heterologous protein production include *Escherichia coli, Saccharomyces cerevisiae,* insect cell expression systems and various mammalian systems. Different host systems that are used for heterologous protein expression have their own advantages and disadvantages as in the case of bacterial systems, a good amount of protein is produced but the quality of the protein may not be appropriate. Similarly, eukaryotic hosts also suffer from specific merits and demerits. Different promoter systems that are used for gene expression also influence protein production in case of both prokaryotic and eukaryotic hosts. The selection of the appropriate host depends on the amount of the protein, quality of the protein in terms of functional activity and its application. In this chapter, we will discuss the principle of heterologous gene expression, different types of hosts organism available for heterologous protein production. We will also present the current developments and modifications which are being done in various host systems to improve the process of protein production. We will elaborate on the different types of vectors that are used for recombinant protein production, their contrasting features and different promoter systems used in the vectors, importance or promoters with respect to protein production. In addition to this, we will also elaborate on different aspects related to the overexpression and purification of recombinant heterologous protein and the type of tags and chromatographic techniques used for the purification of different types of proteins.

Keywords: Affinity tags, Expression hosts, Heterologous protein production, Over expression, Protein production, Strong promoters.

INTRODUCTION

Recombinant protein production is becoming extremely important in the present

[*] **Corresponding author Sanket Kaushik:** Amity Institute of Biotechnology, Amity University Rajasthan, Amity Education Valley, Kant Kalwar, NH-11C, Jaipur-Delhi Highway, Jaipur, India; Tel: 01426-405678; E-mail: skaushik@jpr.amity.edu

era for providing an adequate amount of important proteins in research and medicine [1 - 3]. Several proteins are used as therapeutic agents for treatment of number diseases including diabetes, hereditary defects, pathogenic infections, cancer and even in AIDS [4, 5]. Some commonly used recombinant proteins are insulin, hormones, antibodies, enzymes and anticoagulants [6 - 8]. Looking at the diverse applications of recombinant proteins, the demand of such proteins is increasing day by day. For the successful production of heterologous protein, selection of an appropriate host is very important. Host cells have a major impact on the quantity and quality of the recombinant protein produced. Different types of host cells are available for successful production of recombinant protein. People are already using bacteria, yeast, insect cell lines, mammalian cell lines for heterologous protein production [9 - 12]. There cannot be a single ideal host for the production of every protein therefore, there is an existence of different types of host cells that are suited for the production of different types of proteins. Once the host cell has been selected then the selection of an appropriate vector becomes easier, which is considered extremely important for recombinant protein production. Generally, plasmids are chosen as a vector in experiments involving recombinant protein production [13]. Plasmids are autonomously replicating extrachromosomal DNA sequences that are characterized by the presence of selectable markers, promoters, polylinkers and fusion protein tags *etc.* [14]. Due to the presence of different types of polylinkers, selectable markers and promoters, it is possible to arrange them in a number of combinations to have different types of plasmids for heterologous protein production. The solubility tag is also a very important component of a plasmid vector. Solubility tags are used to purify the recombinantly produced protein by a simple method. Apart from protein purification, solubility tags can also help in enhancing the solubility of some recombinant proteins [15, 16]. The most important and crucial step in heterologous protein expression is the recovery and the purification of the protein of interest. However, this humongous task is made easy by the use of affinity-based chromatographic purification procedures by which the protein of interest can be purified in a single step. Once the protein is purified, then the only important thing is to confirm that the protein is produced in its native functional form. The activity can be measured by different spectroscopic techniques depending upon the specific protein produced [17].

Types of Host Cells Used for Heterologous Protein Production

Several different host cells are used to produce heterologous protein ranging from bacteria to a mammalian cell line. For heterologous expression of all recombinant proteins, none of the hosts are ideal, all the host cells have their own advantages and disadvantages (Table **1**).

Bacterial Expression System

There are several bacterial protein expression systems that can be used in heterologous protein production yet, *E. coli* remains the most widely used expression host. *E. coli* is the most common choice as it is not much expensive, culturing cost is not too much, its genome is well known and genetic manipulation can be done more accurately, easy scale-up is possible, due to its rapid growth rate and most importantly due to the availability of a large number of compatible vectors that can be used in *E. coli* [18]. *E. coli* also has some disadvantages, being a prokaryotic host, it does not have the machinery to perform post-translational modifications. Some of the eukaryotic proteins which are expressed in *E. coli* are not expressed in their functional native state because *E. coli* is not able to perform the appropriate post-translation modifications [19]. Complications are also observed when a protein possesses a lot of disulfide bonds and sometimes when *E. coli* is not able to fold the expressed protein correctly. Such misfolded proteins are expressed in the form of inactive protein aggregates known as inclusion bodies [20]. Although there are various methods available by which these inclusion bodies can be refolded into a fully folded native protein [21, 22]. Most of the proteins which are expressed in *E. coli* are either produced in the cytoplasm or in periplasmic space. Disulfide bond formation is difficult in the cytoplasm due to reducing environment, therefore, such proteins should be synthesized and localized in periplasmic space where the disulfide bond formation is comparatively easier due to its non-reducing environment [23]. Protein expression in *E.coli* can also become a limiting factor due to the deficiency of rare tRNAs in *E.coli* which are present in plenty of amount in the original host for that protein. This is due to the different codon usage bias of different organisms which causes deficit in the corresponding tRNAs in the heterologous host [24, 25]. To overcome this problem, modified strains of *E. coli* BL21 (DE3) cells are used which supply extra copies of rare codon tRNAs to the cell [26]. In line with this, many other modified strains of *E. coli* were developed for more convenient, cost-effective, time efficient recombinant protein production (Table **2**).

Table 1. Types of expression hosts used for heterologous protein production.

S.No.	Organism	Advantages	Disadvantages
1.	**Bacteria:** *Escherichia coli*	• Well established system and easy to manipulate • Safe & inexpensive to grow • Suitable for a variety of labelling *viz.* ^{13}C, ^{15}N, ^{35}S, ^{3}H, Se-Met, *etc.*	• Lack of eukaryotic PTM *i.e.* glycosylation, lipidation, *etc.* • Often the problem with solubility with proteins • Eukaryotic chaperones missing • Expression to protein folding ratio is too high
	Bacillus subtilis	• High expression rate • Direct extracellular protein expression • Improve the quality and quantity of the secreted foreign proteins such as interferon, growth hormone, pepsinogen and epidermal growth factor.	• High secretion of extracellular proteases • Lacks well-regulated inducible vectors

(Table 1) cont.....

S.No.	Organism	Advantages	Disadvantages
	Pseudomonas putida	• Facilitates natural product biosynthesis such as rhamnolipids, terpenoids. • Tolerance to xenobiotics. • Suitable for the production of antimicrobial agents.	• Continuous high expression level can be toxic.
2.	**Fungi:** *Saccharomyces cerevisiae*	• Possess the ability to perform PTM and secretion. • Least cost of post-fermentation *in vitro* purification and modification. • More tolerant to low pH, high sugar and ethanol concentrations, making it suitable for industrial scale fermentations.	• The growth rate is slow as compared to bacterial systems • Hyper glycosylation of expressed proteins.
	Aspergillus nidulans	• Recombinant strains can be cultured in the absence of selection. • Multiple copies of the expression cassette can result in improved levels of expression. • Auxotrophic markers can be used. • Intracellular expression can be achieved at high levels.	
	Pichia pastoris	• High level expression, low cost, and simple culture conditions. • Distinguished production system for its growth to very high cell densities. • Availability of strong and tightly regulated promoters. • Option to produce recombinant proteins either intracellularly or in secretory fashion. • Viable tool for large scale production of integral membrane proteins for structural studies.	• Long incubation time and slow growth rate. • The use of methanol as inducer has certain hazards. • Glycosylation is different from mammalian systems.
3.	**Animal:** Insect cells	• High-level expression. • Faster growth rate. • Effective protein folding and extensive PTM. • Glycosylation is like mammalian cells. • No endotoxins.	• Expensive medium. • Secretion pathway of pro-peptide is inefficient. • Virus infection can lead to cell lysis and protein degradation.
	Mammalian cells	• Higher expression levels. • Suspension culture scalable to large scale. • Full and effective PTM. • Effective protein folding. • Suitable for secreting proteins. • No endotoxins.	• Expensive medium. • Complex growth conditions. • Production is slow.

Table 2. Types of modified *E.coli* expression strains used for heterologous protein production.

S.No.	Strain	Features	Antibiotic resistance	Supplier
1.	BL21	• Widely used non-T7 expression *E. coli* strain. • Ideal for P_{lac}, P_{tac}, $P_{trc}P_{araBAD}$ expression vectors.	Ampicillin	NEB
2.	BL21 (DE3)	• T7 RNA Polymerase under the Control of the lac UV5 Promoter: Inducible protein expression. • Deficient in Proteases Lon and OmpT: Increased stability of the expressed protein.	Ampicillin	NEB
3.	BL21 (DE3) Rosetta	• Designed to enhance the expression of eukaryotic proteins that contain codons rarely used in *E. coli*. • These strains supply tRNAs for AGG, AGA, AUA, CUA, CCC, GGA codons.	Chloramphenicol	Novagen
4.	BL21 (DE3) Star	• Enhanced expression of nontoxic recombinant proteins. • It contains rne131 for the promotion of high mRNA stability and protein yield. • Optimized for use with low copy number, T7 promoter-based plasmids.	Ampicillin	Thermo Scientific
5.	BL21(DE3) pLysS	pLysS Plasmid: Lower background expression of target genes.	Ampicillin	Thermo Scientific
6.	Shuffle T7	• T7 expression • Engineered *E. coli* K12 to promote disulfide bond formation in the cytoplasm • DsbC promotes the correction of mis-oxidized proteins into their correct form.	Ampicillin	NEB
7.	BL21 (DE3) Tuner	• These strains are *lacZY* deletion mutants. • Enables adjustable levels of protein expression. • The lac permease (*lacY*) mutation allows uniform entry of IPTG into all cells in the population, which produces a concentration-dependent, homogeneous level of induction.		Novagen
8.	BL21 (DE3) Codon+	• Engineered to contain extra copies of genes that encode the tRNAs that most frequently limit translation of the heterologous proteins in *E. coli*. • The availability of tRNAs allows high-level expression of many heterologous recombinant genes.	Chloramphenicol	Agilent Technologies

As far as recovery of the recombinant protein is concerned, recombinant proteins expressed in *E. coli* can be easily recovered and purified by various methods. Initially, cells are disrupted by chemical method, enzymatic method or mechanical method which is generally followed by centrifugation or filtration. The protein of interest is purified from the cell pellet obtained after centrifugation using different chromatographic techniques [27 - 30].

In addition to *E.coli*, other prokaryotic host organisms are also being used for heterologous protein production like *Bacillus subtilis*, *Lactococcus lactis* and *Pseudomonas fluorescens*, *etc.* [31 - 33]. *B. subtilis* is regarded as a one of the most widely used expression hosts as it has several advantages like, the proteins produced in *B. subtilis* are directly secreted in the culture media which ease the protein recovery process after expression, it is non-pathogenic bacterium like *E. coli*, chances of formation of inactive protein aggregates are lesser in case of *B. subtilis* as it has the information needed for protein folding, secretion and disulfide bond formation *etc.* [34 - 36]. Most importantly it has lesser codon usage bias as compared to *E.coli* so the rare tRNA concentration is not a major problem with *Bacillus subtilis* [37]. Like every other host, it also suffers from some drawbacks, as it produces extracellular proteins these proteins are degraded by number of extracellular proteases produced by the bacteria itself. Although, now adays modified strains of *B. subtilis* are being developed which are protease-free strains [38]. These strains do not produce extracellular or membrane-bound proteases, so the proteins which are produced recombinantly are safe from proteases activity. *L. lactis* also a promising alternative of expression host as compared to the other prokaryotic host systems. Like other bacterial hosts, it also has a rapid growth rate, it has a strong promoter which allows synthesis of proteins at a higher rate, it does not have proteolytic activity, therefore, the proteins produced are largely safe from any proteolytic attack, it is relatively a better host to produce membrane proteins [39, 40]. On the other hand, *L. lactis* lacks chaperone proteins that are required for correct protein folding and it also lacks gene coding for disulfide isomerase which is required for correct disulfide bond formation [41]. In addition to this *P. fluorescens*, *P. putida*, *Caulobacter crescentus* and *B. megaterium*, *etc.* are also used as host strains for recombinant protein production.

Fungi as Expression System

Different fungal systems are also available for recombinant protein production which includes *Saccharomyces cerevisiae*, *Aspergillus nidulans* and *Pichia pastoris* [42, 43]. All of them offer a range of features that can be used in protein expression. *S. cerevisiae* is a very important and highly used expression system, apart from being a model organism for cell biology and genetics studies [44]. *S.*

cerevisiae is a unique vector as it has advantages of both eukaryotic and prokaryotic expression systems. Like a prokaryotic expression system, it also employs inexpensive culture media, it is very easy to grow, like *E.coli*, it also has a higher growth rate, *etc*. Being a eukaryotic host, it also can carry out post-translation modifications, proper protein folding and other protein modifications [45]. On the other hand, the proteins expressed in *S. cerevisiae* are usually hyperglycosylated, which may affect the functional activity of the protein. However, modified host strains are being used in which the activity of Golgi mannosyltransferase has been suppressed. Suppressing the activity of these enzymes will block hyper glycosylation in the recombinantly expressed proteins [46].

A. nidulans shares all the advantages with *S. cerevisiae* for protein expression still its use as an expression host is not very promising [47, 48]. It can be attributable to the fact that *A. nidulans* has a tendency to produce toxins which can harm the expression of the protein. To overcome these genetically modified strains are being made which has an incomplete or missing pathway of toxin synthesis [49, 50]. Also, the transformation procedures in *A. nidulans* are not that easy as compared to *E.coli* and *S. cerevisiae*. Furthermore, many convenient procedures for transformation of a foreign gene in *A. nidulans* are in use now [51].

In recent years *P. pastoris* is being used extensively for recombinant protein expression [52]. It has an additional advantage as compared to other fungi-based hosts as it is able to glycosylate proteins like the animals [53]. Glycosylation pattern is not exactly like animals but it is better than *S. cerevisiae*, where the proteins are hyperglycosylated. To add to this, recombinant proteins can also be directly secreted in the culture media or in other words proteins can be synthesized extra cellularly in *P. pastoris* by which the recovery of the protein becomes easier.

Animal Based Expression Systems

Insect-based hosts cells and mammalian cells are widely used these days [54, 55]. Although culturing cost is very high in these host systems and it has a very slow growth rate, insect cells have several characters that make them a better host for recombinant protein production. Proteins produced in insect cells undergo appropriate post-translation modifications, so the chances of getting a fully functional protein are much higher in case of insect cell lines. Insect cell-based hosts particularly baculovirus based systems offer a powerful expression and efficient system for producing high levels of recombinant protein expression [56]. Baculovirus has a host range limited to some specific invertebrates hence, it is non-infectious to vertebrates and safe to work with most mammalian viruses.

Proteins are properly glycosylated in insect cells, but the glycosylation pattern is not identical to mammals. If glycosylation is critical for the activity of the protein than in that case, there are chances that the protein will not function properly [57].

Mammalian systems are ideal for protein expression as far as human proteins are concerned [58, 59]. Proteins produced in a mammalian host possess proper protein folding, post-translation modifications, signal peptide synthesis and proper glycosylation of the protein, endotoxin free, *etc*. It has a few drawbacks like it is very expensive, as the cost of cultivation is too high and it has a very slow growth rate [60]. Before selecting an ideal expression host, cost and time are major concerns. Prokaryotic hosts provide good quantity of protein and eukaryotic hosts provide good quality of protein. None of the hosts are ideal for expression of every protein. The selection of host is dependent on the nature of the protein which must be expressed recombinantly.

Importance of Vectors in Heterologous Gene Expression

The choice of suitable expression vector is very important for successful protein production. There are several components in an expression vector which can influence the rate and the amount of protein produced [61]. The most important component of an expression vector which influences the amount of protein production is the promoter. In prokaryotic expression system, usually inducible promoters are used which are lac operator and T7 promoter based [62]. T7 promoter is a strong promoter that is derived from bacteriophage. T7 RNA polymerase gene is usually inserted in all the engineered hosts of prokaryotic origin for the activation of T7 polymerase. Interestingly, this T7 RNA polymerase gene is under the control of lac operator in the bacterial genome [63]. As soon as the inducers are added in the media, lac operator gets activated and produces T7 RNA polymerase which in turn activates the T7 promoter in the expression vector and the protein of interest is overexpressed (Fig. **1**).

In contrast, most of the vectors used in the yeast system are shuttle vectors. Shuttle vectors are the vectors that can be used in two different hosts as they have two origins of replications, one for prokaryotic system and one for the eukaryotic system [64]. As compared to the prokaryotic system where the selection of the recombinant is done based on antibiotic resistance gene selection, in yeast it is done based on auxotrophic markers [65]. Yeast host systems are auxotrophic for uracil or different amino acids are used with plasmids harboring the auxotrophic markers copy of the gene. The selection of the recombinant is done based on the colonies surviving on an auxotrophic culture media as the recombinant plasmids provide the auxotrophic copy of the gene. Although, negative selection with antibiotic resistance has also been reported in the yeast system where negative selection is done. Promoters have a major influence on the amount of protein produced. Promoters in yeast are constitutive, strong promoters are derived from major metabolic pathways like glycolytic pathway (P_{PGK1}, P_{TDH3}, P_{ENO2}, and P_{TPI1}), galactose metabolic pathway ($P_{GAL10/GAL1}$), chaperone promoters (P_{SSA1} and P_{SSB1}) and translational elongation factor promoters (*TEF:* P_{TEF1}, P_{TEF2} and P_{YEF3}), *etc.* [66].

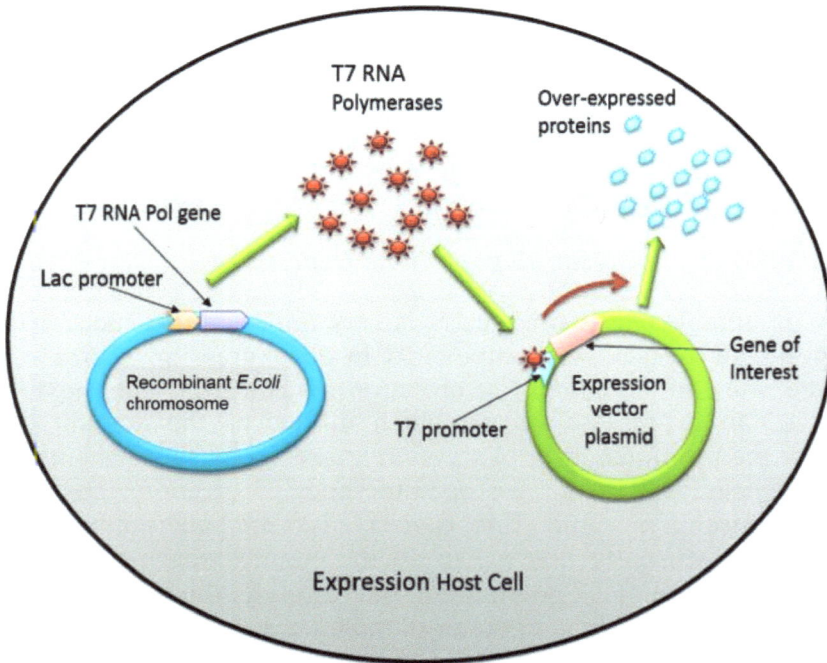

Fig. (1). Mechanism of T7 Promoter activation.

Baculoviral Expression Vector System is a widely used eukaryotic expression system for heterologous protein production in insect cells [67]. It has several contrasting features which make it a preferred expression vector for expression of different proteins. Baculoviral Expression Vector System has now become easy to use with the recovery of recombinant protein produced within five days. Several genes can be expressed simultaneously or multimeric proteins can be expressed and assembled successfully in the Baculoviral Expression Vector System. There is no limit to the size of the recombinant protein produced as we have a limitation in the size of protein in bacterial systems. Most importantly the recombinant proteins produced are fully functional with proper post-translation modifications like glycosylation, phosphorylation, acetylation, *etc.* Additionally, recombinant proteins produced are localized properly in the exact cellular compartment as the native host and several affinity-based purification procedures have been developed for simple purification of the recombinant protein. Several mammalians based viral-based and non-viral based expression vectors are also available for protein production [68].

Purification of Recombinant Protein

Purification of recombinantly produced proteins can be made highly efficient by adding a known sequence in the vector known as tag. Tag based purification is well known for recombinant protein purification in the prokaryotic and the eukaryotic expression systems [69]. Tags are of two types: affinity tag and solubility enhancing tag. Affinity tag only helps in single step purification of the protein of interest while solubility enhancing tags not only helps in protein purification, but also help in proper solubilization of the recombinant proteins where the formation of inclusion body is a major issue [70, 71]. Commonly used affinity tag and solubility enhancing tags are poly-histidine tag and "Glutathione *S-* transferase" (GST) tag respectively [72, 73]. Both the tags can be used in any expression system whether prokaryotic or eukaryotic. Small poly-histidine tags provide less metabolic burden to the host while GST tag provides a heavy metabolic burden to the host as it is large. GST tag can enhance the solubility of the recombinant protein, while poly-histidine does not have any effect of enhancing the solubility of the recombinant protein. Poly-histidine tag may or may not be removed after protein purification as being small, it does not affect the activity of the recombinant protein. On the other hand, GST tag must be removed after the purification of the recombinant protein as it large and if it is not removed from the recombinant protein it may affect the activity of the recombinant protein. It is also reported that a poly-histidine tag being small does not hinder the crystallization, while GST being large may interfere with the crystallization of the purified protein [74, 75]. Affinity tag or solubility enhancing tag has several advantages and disadvantages listed in Table **3**.

Table 3. Comparison between Poly-Histidine tag and GST tag.

S.No.	Poly-histidine Tag	GST Tag
01.	Poly-histidine tag is a six histidine (His) motif fused to recombinant proteins that are often present at the N or C-terminus of the protein.	Glutathione *S*-transferase (GST) is a 26 KDa protein often fused to recombinant proteins at the N or C-terminus region.
02.	It is also known as hexa histidine-tag, 6xHis-tag, His6 tag or simply His-tag.	It is commonly known as Glutathione *S*-transferase tag or simply GST tag.
03.	It was invented by Roche.	It was invented by D.B. Smith.

(Table 3) cont.....

S.No.	Poly-histidine Tag	GST Tag
04.	His-tags are often used for affinity purification of recombinant proteins.	GST tags are often used for enhancing solubility and affinity purification of recombinant proteins.
05.	His-tag is exposed on the protein surface.	GST tags are fused to recombinant proteins and enhance its solubilization.
06.	It can be easily cleaved from the protein after protein purification, but removal is not necessary.	It can be easily cleaved from the protein after protein purification and removal may be necessary.
07.	Ni-NTA Agarose is a nickel-charged affinity resin that can be used to purify recombinant proteins containing a His-tag.	Sepharose embedded glutathione is used to purify recombinant proteins containing GST tags.
08.	The technique used to purify His-tag protein is known as Immobilized Metal Affinity Chromatography (IMAC), which is based on the interaction between histidine residues and divalent metal ions immobilized on resins.	Affinity chromatography is the basic technique, where glutathione is immobilized on a matrix (basically sepharose) and interacts with GST tags.
09.	Imidazole is used to elute His-tagged proteins.	Reduced glutathione is used to elute GST-tagged proteins.
10.	The oligopeptide (His tag) too small to have impact on native state.	A drawback of this assay is that the protein of interest is attached to GST, altering its native state.
11.	Most preferred tag to purify recombinant proteins.	Preferred when solubility enhancement of protein is required.

Applications of Heterologous Protein Production

1. ***Production of Human Therapeutic Agents:*** Heterologous protein expression is used to overexpress many important proteins like insulin, epidermal growth factor, proinsulin, fibroblast growth factor, human coagulation factor, *etc.* These proteins are used for treatment of several diseases like diabetes, diabetic foot ulcers, cardiovascular diseases, *etc.* Heterologous production is done to get more yield of a protein than the original host by exploiting different expression vectors and enzymes.
2. ***Production of Important Vaccines:*** Different important vaccines are produced in heterologous hosts. Inactivated or live attenuated virus are being used for vaccination but due to their number of shortcomings, recombinant protein vaccines have gained interest in recent years. Recombinant protein vaccines have a well-defined structure as they are made up of small fraction of microbial component which is produced recombinantly. These vaccines are safer than the conventional vaccines as they only contain a fraction of a recombinant protein which is not infectious.
3. ***Diagnosis of Infections:*** Several important diseases including HIV infections

are diagnosed by ELISA-based methods involving use of recombinantly produced HIV proteins which are used to detect antibodies that are produced by the body in response to HIV infections. Such new methods have improved the accuracy and reliability of disease diagnostics.

4. ***Use in Basic Research:*** Heterologous protein production had made possible availability of ample amount of respective protein for detailed molecular and structural studies. Such studies are essential to elucidate the detailed functional mechanism of a protein which can further make possible design of several biophysical and biochemical studies leading to potential lead development and drug discovery.

5. ***Use in Biotechnology Industries:*** Recombinantly produced proteins have important contributions in the field of biotechnology including agricultural biotechnology, food biotechnology and bioengineering. Many important proteins are added as supplements in the food to increase the nutritional value of the food or in the agricultural industry to increase the animal performance and the feed nutritional value. Chymosin is a common example of recombinantly produced enzyme used as a food additive.

6. ***Production of Important Peptides:*** Several important peptides are known to have antibacterial, antiviral, antifungal or antitumor properties. Such peptides are available in nature in a limited amount. Such peptides are produced heterologously in excess which can be used in various fields in biotechnology, ranging from medical biotechnology, food biotechnology and industrial biotechnology.

CONSENT FOR PUBLICATION

Not applicable.

CONFLICT OF INTEREST

The author confirms that this chapter contents have no conflict of interest.

ACKNOWLEDGEMENTS

Declared none.

REFERENCES

[1]   Baneyx F. Recombinant protein expression in *Escherichia coli*. Curr Opin Biotechnol 1999; 10(5): 411-21.
[http://dx.doi.org/10.1016/S0958-1669(99)00003-8] [PMID: 10508629]

[2]   Wurm FM. Production of recombinant protein therapeutics in cultivated mammalian cells. Nat Biotechnol 2004; 22(11): 1393-8.
[http://dx.doi.org/10.1038/nbt1026] [PMID: 15529164]

[3]   Davis TR, Wickham TJ, McKenna KA, Granados RR, Shuler ML, Wood HA. Comparative

recombinant protein production of eight insect cell lines *in vitro*. Cell Dev Biol Anim 1993; 29A(5): 388-90.
[http://dx.doi.org/10.1007/BF02633986] [PMID: 8314732]

[4] Buckel F. Recombinant proteins for therapy. Trends Pharmacol Sci 1996; 17(12): 450-6.
[http://dx.doi.org/10.1016/S0165-6147(96)01011-5] [PMID: 9014499]

[5] Andersen DC, Krummen L. Recombinant protein expression for therapeutic applications. Curr Opin Biotechnol 2002; 13(2): 117-23.
[http://dx.doi.org/10.1016/S0958-1669(02)00300-2] [PMID: 11950561]

[6] Hanahan D. Heritable formation of pancreatic β-cell tumours in transgenic mice expressing recombinant insulin/simian virus 40 oncogenes. Nature 1985; 315(6015): 115-22.
[http://dx.doi.org/10.1038/315115a0] [PMID: 2986015]

[7] Jensen EB, Carlsen S. Production of recombinant human growth hormone in *Escherichia coli*: expression of different precursors and physiological effects of glucose, acetate, and salts. Biotechnol Bioeng 1990; 36(1): 1-11.
[http://dx.doi.org/10.1002/bit.260360102] [PMID: 18592603]

[8] Van Berkel PHC, Bout A, Logtenberg T, *et al.* US Patent No 7,262,028. Washington, DC: U.S. Patent and Trademark Office. 2007.

[9] Miroux B, Walker JE. Over-production of proteins in *Escherichia coli*: mutant hosts that allow synthesis of some membrane proteins and globular proteins at high levels. J Mol Biol 1996; 260(3): 289-98.
[http://dx.doi.org/10.1006/jmbi.1996.0399] [PMID: 8757792]

[10] Punt PJ, van Biezen N, Conesa A, Albers A, Mangnus J, van den Hondel C. Filamentous fungi as cell factories for heterologous protein production. Trends Biotechnol 2002; 20(5): 200-6.
[http://dx.doi.org/10.1016/S0167-7799(02)01933-9] [PMID: 11943375]

[11] Maiorella B, Inlow D, Shauger A, *et al.* Large-scale insect cell-culture for recombinant protein production. Nat Biotechnol 1988; 6(12): 1406.
[http://dx.doi.org/10.1038/nbt1288-1406]

[12] Demain AL, Vaishnav P. Production of recombinant proteins by microbes and higher organisms. Biotechnol Adv 2009; 27(3): 297-306.
[http://dx.doi.org/10.1016/j.biotechadv.2009.01.008] [PMID: 19500547]

[13] di Guan C, Li P, Riggs PD, *et al.* Vectors that facilitate the expression and purification of foreign peptides in *E.coli* by fusion to maltose-binding protein (Recombinant DNA; plasmids; cross-linked amylose affinity chromatography; starch). Gene 1988; 67(1988): 21-30.
[http://dx.doi.org/10.1016/0378-1119(88)90004-2] [PMID: 2843437]

[14] Thomas CM, Summers D. plasmids e LS. 2001.

[15] Esposito D, Chatterjee DK. Enhancement of soluble protein expression through the use of fusion tags. Curr Opin Biotechnol 2006; 17(4): 353-8.
[http://dx.doi.org/10.1016/j.copbio.2006.06.003] [PMID: 16781139]

[16] Lichty JJ, Malecki JL, Agnew HD, Michelson-Horowitz DJ, Tan S. Comparison of affinity tags for protein purification. Protein Expr Purif 2005; 41(1): 98-105.
[http://dx.doi.org/10.1016/j.pep.2005.01.019] [PMID: 15802226]

[17] Liu F, Zhang F, Jin Z, *et al.* Determination of acetolactate synthase activity and protein content of oilseed rape (Brassica napus L.) leaves using visible/near-infrared spectroscopy. Anal Chim Acta 2008; 629(1-2): 56-65.
[http://dx.doi.org/10.1016/j.aca.2008.09.027] [PMID: 18940321]

[18] Baneyx F. Recombinant protein expression in *Escherichia coli*. Curr Opin Biotechnol 1999; 10(5): 411-21.
[http://dx.doi.org/10.1016/S0958-1669(99)00003-8] [PMID: 10508629]

[19]   Georgiou G, Valax P. Expression of correctly folded proteins in *Escherichia coli*. Curr Opin Biotechnol 1996; 7(2): 190-7.
[http://dx.doi.org/10.1016/S0958-1669(96)80012-7] [PMID: 8791338]

[20]   Kane JF, Hartley DL. Formation of recombinant protein inclusion bodies in *Escherichia coli*. Trends Biotechnol 1988; 6(5): 95-101.
[http://dx.doi.org/10.1016/0167-7799(88)90065-0]

[21]   Lilie H, Schwarz E, Rudolph R. Advances in refolding of proteins produced in *E. coli*. Curr Opin Biotechnol 1998; 9(5): 497-501.
[http://dx.doi.org/10.1016/S0958-1669(98)80035-9] [PMID: 9821278]

[22]   Vallejo LF, Rinas U. Strategies for the recovery of active proteins through refolding of bacterial inclusion body proteins. Microb Cell Fact 2004; 3(1): 11.
[http://dx.doi.org/10.1186/1475-2859-3-11] [PMID: 15345063]

[23]   Choi JH, Lee SY. Secretory and extracellular production of recombinant proteins using *Escherichia coli*. Appl Microbiol Biotechnol 2004; 64(5): 625-35.
[http://dx.doi.org/10.1007/s00253-004-1559-9] [PMID: 14966662]

[24]   Kane JF. Effects of rare codon clusters on high-level expression of heterologous proteins in *Escherichia coli*. Curr Opin Biotechnol 1995; 6(5): 494-500.
[http://dx.doi.org/10.1016/0958-1669(95)80082-4] [PMID: 7579660]

[25]   Gustafsson C, Govindarajan S, Minshull J. Codon bias and heterologous protein expression. Trends Biotechnol 2004; 22(7): 346-53.
[http://dx.doi.org/10.1016/j.tibtech.2004.04.006] [PMID: 15245907]

[26]   Rosano GL, Ceccarelli EA. Recombinant protein expression in *Escherichia coli*: advances and challenges. Front Microbiol 2014; 5: 172.
[http://dx.doi.org/10.3389/fmicb.2014.00172] [PMID: 24860555]

[27]   Johnson BH, Hecht MH. Recombinant proteins can be isolated from *E. coli* cells by repeated cycles of freezing and thawing. Biotechnology (N Y) 1994; 12(13): 1357-60.
[PMID: 7765566]

[28]   Gabor EM, de Vries EJ, Janssen DB. Efficient recovery of environmental DNA for expression cloning by indirect extraction methods. FEMS Microbiol Ecol 2003; 44(2): 153-63.
[http://dx.doi.org/10.1016/S0168-6496(02)00462-2] [PMID: 19719633]

[29]   Smith DB, Johnson KS. Single-step purification of polypeptides expressed in *Escherichia coli* as fusions with glutathione S-transferase. Gene 1988; 67(1): 31-40.
[http://dx.doi.org/10.1016/0378-1119(88)90005-4] [PMID: 3047011]

[30]   Guan KL, Dixon JE. Eukaryotic proteins expressed in *Escherichia coli*: an improved thrombin cleavage and purification procedure of fusion proteins with glutathione S-transferase. Anal Biochem 1991; 192(2): 262-7.
[http://dx.doi.org/10.1016/0003-2697(91)90534-Z] [PMID: 1852137]

[31]   Wu XC, Lee W, Tran L, Wong SL. Engineering a Bacillus subtilis expression-secretion system with a strain deficient in six extracellular proteases. J Bacteriol 1991; 173(16): 4952-8.
[http://dx.doi.org/10.1128/jb.173.16.4952-4958.1991] [PMID: 1907264]

[32]   Kunji ER, Slotboom DJ, Poolman B. Lactococcus lactis as host for overproduction of functional membrane proteins. Biochim Biophys Acta 2003; 1610(1): 97-108.
[http://dx.doi.org/10.1016/S0005-2736(02)00712-5] [PMID: 12586384]

[33]   Olsen RH, Shipley P. Host range and properties of the Pseudomonas aeruginosa R factor R1822. J Bacteriol 1973; 113(2): 772-80.
[PMID: 4632321]

[34]   Westers L, Westers H, Quax WJ. Bacillus subtilis as cell factory for pharmaceutical proteins: a

biotechnological approach to optimize the host organism. Biochim Biophys Acta 2004; 1694(1-3): 299-310.
[http://dx.doi.org/10.1016/j.bbamcr.2004.02.011] [PMID: 15546673]

[35]   Wong SL. Advances in the use of Bacillus subtilis for the expression and secretion of heterologous proteins. Curr Opin Biotechnol 1995; 6(5): 517-22.
[http://dx.doi.org/10.1016/0958-1669(95)80085-9] [PMID: 7579663]

[36]   Wu XC, Ng SC, Near RI, Wong SL. Efficient production of a functional single-chain antidigoxin antibody *via* an engineered Bacillus subtilis expression-secretion system. Biotechnology (N Y) 1993; 11(1): 71-6.
[PMID: 7763487]

[37]   Sharp PM, Cowe E, Higgins DG, Shields DC, Wolfe KH, Wright F. Codon usage patterns in *Escherichia coli*, Bacillus subtilis, Saccharomyces cerevisiae, Schizosaccharomyces pombe, Drosophila melanogaster and Homo sapiens; a review of the considerable within-species diversity. Nucleic Acids Res 1988; 16(17): 8207-11.
[http://dx.doi.org/10.1093/nar/16.17.8207] [PMID: 3138659]

[38]   Wu XC, Lee W, Tran L, Wong SL. Engineering a Bacillus subtilis expression-secretion system with a strain deficient in six extracellular proteases. J Bacteriol 1991; 173(16): 4952-8.
[http://dx.doi.org/10.1128/jb.173.16.4952-4958.1991] [PMID: 1907264]

[39]   Kunji ER, Slotboom DJ, Poolman B. Lactococcus lactis as host for overproduction of functional membrane proteins. Biochim Biophys Acta 2003; 1610(1): 97-108.
[http://dx.doi.org/10.1016/S0005-2736(02)00712-5] [PMID: 12586384]

[40]   Wells JM, Wilson PW, Norton PM, Gasson MJ, Le Page RW. Lactococcus lactis: high-level expression of tetanus toxin fragment C and protection against lethal challenge. Mol Microbiol 1993; 8(6): 1155-62.
[http://dx.doi.org/10.1111/j.1365-2958.1993.tb01660.x] [PMID: 8361360]

[41]   Drouault S, Anba J, Bonneau S, Bolotin A, Ehrlich SD, Renault P. The peptidyl-prolyl isomerase motif is lacking in PmpA, the PrsA-like protein involved in the secretion machinery of Lactococcus lactis. Appl Environ Microbiol 2002; 68(8): 3932-42.
[http://dx.doi.org/10.1128/AEM.68.8.3932-3942.2002] [PMID: 12147493]

[42]   Nevalainen KM, Te'o VS, Bergquist PL. Heterologous protein expression in filamentous fungi. Trends Biotechnol 2005; 23(9): 468-74.
[http://dx.doi.org/10.1016/j.tibtech.2005.06.002] [PMID: 15967521]

[43]   Punt PJ, van Biezen N, Conesa A, Albers A, Mangnus J, van den Hondel C. Filamentous fungi as cell factories for heterologous protein production. Trends Biotechnol 2002; 20(5): 200-6.
[http://dx.doi.org/10.1016/S0167-7799(02)01933-9] [PMID: 11943375]

[44]   Ghaemmaghami S, Huh WK, Bower K, *et al.* Global analysis of protein expression in yeast. Nature 2003; 425(6959): 737-41.
[http://dx.doi.org/10.1038/nature02046] [PMID: 14562106]

[45]   Gomes AR, Byregowda SM, Veeregowda BM, *et al.* An overview of heterologous expression host systems to produce recombinant proteins. Adv Anim Vet Sci 2016; 4(7): 346-56.
[http://dx.doi.org/10.14737/journal.aavs/2016/4.7.346.356]

[46]   Tang H, Wang S, Wang J, *et al.* N-hypermannose glycosylation disruption enhances recombinant protein production by regulating secretory pathway and cell wall integrity in Saccharomyces cerevisiae. Sci Rep 2016; 6: 25654.
[http://dx.doi.org/10.1038/srep25654] [PMID: 27156860]

[47]   Devchand M, Gwynne DI. Expression of heterologous proteins in Aspergillus. J Biotechnol 1991; 17(1): 3-9.
[http://dx.doi.org/10.1016/0168-1656(91)90022-N] [PMID: 1367014]

[48]   Lubertozzi D, Keasling JD. Developing Aspergillus as a host for heterologous expression. Biotechnol Adv 2009; 27(1): 53-75.
[http://dx.doi.org/10.1016/j.biotechadv.2008.09.001] [PMID: 18840517]

[49]   Matsushima K, Chang PK, Yu J, Abe K, Bhatnagar D, Cleveland TE. Pre-termination in aflR of Aspergillus sojae inhibits aflatoxin biosynthesis. Appl Microbiol Biotechnol 2001; 55(5): 585-9.
[http://dx.doi.org/10.1007/s002530100607] [PMID: 11414325]

[50]   Matsushima K, Yashiro K, Hanya Y, Abe K, Yabe K, Hamasaki T. Absence of aflatoxin biosynthesis in koji mold (Aspergillus sojae). Appl Microbiol Biotechnol 2001; 55(6): 771-6.
[http://dx.doi.org/10.1007/s002530000524] [PMID: 11525627]

[51]   Campbell EI, Unkles SE, Macro JA, van den Hondel C, Contreras R, Kinghorn JR. Improved transformation efficiency of Aspergillus niger using the homologous niaD gene for nitrate reductase. Curr Genet 1989; 16(1): 53-6.
[http://dx.doi.org/10.1007/BF00411084] [PMID: 2791035]

[52]   Cereghino JL, Cregg JM. Heterologous protein expression in the methylotrophic yeast Pichia pastoris. FEMS Microbiol Rev 2000; 24(1): 45-66.
[http://dx.doi.org/10.1111/j.1574-6976.2000.tb00532.x] [PMID: 10640598]

[53]   Macauley-Patrick S, Fazenda ML, McNeil B, Harvey LM. Heterologous protein production using the Pichia pastoris expression system. Yeast 2005; 22(4): 249-70.
[http://dx.doi.org/10.1002/yea.1208] [PMID: 15704221]

[54]   Barnes LM, Bentley CM, Dickson AJ. Advances in animal cell recombinant protein production: GS-NS0 expression system. Cytotechnology 2000; 32(2): 109-23.
[http://dx.doi.org/10.1023/A:1008170710003] [PMID: 19002973]

[55]   Liljeström P, Garoff H. A new generation of animal cell expression vectors based on the Semliki Forest virus replicon. Biotechnology (N Y) 1991; 9(12): 1356-61.
[http://dx.doi.org/10.1038/nbt1291-1356] [PMID: 1370252]

[56]   Kost TA, Condreay JP, Jarvis DL. Baculovirus as versatile vectors for protein expression in insect and mammalian cells. Nat Biotechnol 2005; 23(5): 567-75.
[http://dx.doi.org/10.1038/nbt1095] [PMID: 15877075]

[57]   Jarvis DL, Finn EE. Modifying the insect cell N-glycosylation pathway with immediate early baculovirus expression vectors. Nat Biotechnol 1996; 14(10): 1288-92.
[http://dx.doi.org/10.1038/nbt1096-1288] [PMID: 9631095]

[58]   Aricescu AR, Lu W, Jones EY. A time- and cost-efficient system for high-level protein production in mammalian cells. Acta Crystallogr D Biol Crystallogr 2006; 62(Pt 10): 1243-50.
[http://dx.doi.org/10.1107/S0907444906029799] [PMID: 17001101]

[59]   Tian Q, Stepaniants SB, Mao M, *et al.* Integrated genomic and proteomic analyses of gene expression in Mammalian cells. Mol Cell Proteomics 2004; 3(10): 960-9.
[http://dx.doi.org/10.1074/mcp.M400055-MCP200] [PMID: 15238602]

[60]   Khan KH. Gene expression in Mammalian cells and its applications. Adv Pharm Bull 2013; 3(2): 257-63.
[PMID: 24312845]

[61]   Gräslund S, Nordlund P, Weigelt J, *et al.* Structural Genomics Consortium China Structural Genomics Consortium Northeast Structural Genomics Consortium. Protein production and purification. Nat Methods 2008; 5(2): 135-46.
[http://dx.doi.org/10.1038/nmeth.f.202] [PMID: 18235434]

[62]   Baneyx F. Recombinant protein expression in *Escherichia coli*. Curr Opin Biotechnol 1999; 10(5): 411-21.
[http://dx.doi.org/10.1016/S0958-1669(99)00003-8] [PMID: 10508629]

[63]   Alexander WA, Moss B, Fuerst TR. Regulated expression of foreign genes in vaccinia virus under the control of bacteriophage T7 RNA polymerase and the *Escherichia coli* lac repressor. J Virol 1992; 66(5): 2934-42.
[PMID: 1560532]

[64]   Christianson TW, Sikorski RS, Dante M, Shero JH, Hieter P. Multifunctional yeast high-copy-number shuttle vectors. Gene 1992; 110(1): 119-22.
[http://dx.doi.org/10.1016/0378-1119(92)90454-W] [PMID: 1544568]

[65]   Boeke JD, LaCroute F, Fink GR. A positive selection for mutants lacking orotidine-5'-phosphate decarboxylase activity in yeast: 5-fluoro-orotic acid resistance. Mol Gen Genet 1984; 197(2): 345-6.
[http://dx.doi.org/10.1007/BF00330984] [PMID: 6394957]

[66]   Hauf J, Zimmermann FK, Müller S. Simultaneous genomic overexpression of seven glycolytic enzymes in the yeast Saccharomyces cerevisiae. Enzyme Microb Technol 2000; 26(9-10): 688-98.
[http://dx.doi.org/10.1016/S0141-0229(00)00160-5] [PMID: 10862874]

[67]   Kost TA, Condreay JP, Jarvis DL. Baculovirus as versatile vectors for protein expression in insect and mammalian cells. Nat Biotechnol 2005; 23(5): 567-75.
[http://dx.doi.org/10.1038/nbt1095] [PMID: 15877075]

[68]   Wurm F, Bernard A. Large-scale transient expression in mammalian cells for recombinant protein production. Curr Opin Biotechnol 1999; 10(2): 156-9.
[http://dx.doi.org/10.1016/S0958-1669(99)80027-5] [PMID: 10209142]

[69]   Zhao X, Li G, Liang S. Several affinity tags commonly used in chromatographic purification. Journal of analytical methods in chemistry 2013; 2013
[http://dx.doi.org/10.1155/2013/581093]

[70]   Arnau J, Lauritzen C, Petersen GE, Pedersen J. Current strategies for the use of affinity tags and tag removal for the purification of recombinant proteins. Protein Expr Purif 2006; 48(1): 1-13.
[http://dx.doi.org/10.1016/j.pep.2005.12.002] [PMID: 16427311]

[71]   Esposito D, Chatterjee DK. Enhancement of soluble protein expression through the use of fusion tags. Curr Opin Biotechnol 2006; 17(4): 353-8.
[http://dx.doi.org/10.1016/j.copbio.2006.06.003] [PMID: 16781139]

[72]   Bornhorst JA, Falke JJ. [16] Purification of proteins using poly-histidineaffinity tags. Methods in enzymology 2000 Jan 1; 326: 245-54.

[73]   Scheich C, Sievert V, Büssow K. An automated method for high-throughput protein purification applied to a comparison of His-tag and GST-tag affinity chromatography. BMC Biotechnol 2003; 3(1): 12.
[http://dx.doi.org/10.1186/1472-6750-3-12] [PMID: 12885298]

[74]   Lichty JJ, Malecki JL, Agnew HD, Michelson-Horowitz DJ, Tan S. Comparison of affinity tags for protein purification. Protein Expr Purif 2005; 41(1): 98-105.
[http://dx.doi.org/10.1016/j.pep.2005.01.019] [PMID: 15802226]

[75]   Scheich C, Sievert V, Büssow K. An automated method for high-throughput protein purification applied to a comparison of His-tag and GST-tag affinity chromatography. BMC Biotechnol 2003; 3(1): 12.
[http://dx.doi.org/10.1186/1472-6750-3-12] [PMID: 12885298]

CHAPTER 6

In Silico Modeling and Drug Designing

Gauri Misra[*] and **Neetu Jabalia**

Amity Institute of Biotechnology, Amity University, Noida, India

Abstract: Rational drug designing encompasses several theoretical methods and *in silico* approaches involving molecular modeling, docking *etc.* to study the behavior and the properties of molecular systems. Specifically, the techniques employed in the fields of computational chemistry, computational biology, nanotechnology, and material science vary in complexity, and theoretical observations depend on the system type and the system size being investigated. Molecular modeling involves both quantum and molecular mechanics (QM & MM). The fundamental concepts of molecular modelling with suitable examples are elaborated. The present chapter also highlights the sequential flow of *in silico* approaches used for the purpose of drug designing, giving a brief snapshot of the software used for this purpose. The advanced techniques and applications of computational methods used for new drug development are explained. Thus, this chapter provides an insight into the various dimensions of computer-aided drug design; its driving force, recent development, and future prospects.

Keywords: Molecular dynamics, Open source drug discovery, QSAR, Structure based drug designing, Virtual screening, Virtual screening.

INTRODUCTION

The drug discovery process is a long process in terms of time duration and resources required. Drug activity is a concomitant result of multiple factors that necessitate the study of its bioavailability, toxicity and metabolism. A plethora of information is generated experimentally that helps in the field of biology and computer sciences. It is effectively used for *in silico* drug designing processes to streamline the drug discovery efforts with an emphasis on lead generation and optimization [1]. Computer assisted drug designing (CADD) processes involve both structure based drug design (SBDD) and ligand based drug design (LBDD) [2]. It reduces the wastage of time and research resources leading to the development of effective drugs. The computational methods include molecular modeling, virtual screening, docking, ligand interaction and molecular dynamics (Fig. **1**). This not only helps in determining the physicochemical properties of

---

[*] **Corresponding author Gauri Misra:** Amity Institute of Biotechnology, Amity University, Noida, India; Tel: +91-9891203994; E-mail: kamgauri@gmail.com

Anupam Jyoti & Neetu Mishra (Eds.)

compounds but also aids in the determination of pharmacokinetic and pharmacodynamic properties of drugs, structural activity relationship (SAR) between ligand and its target.

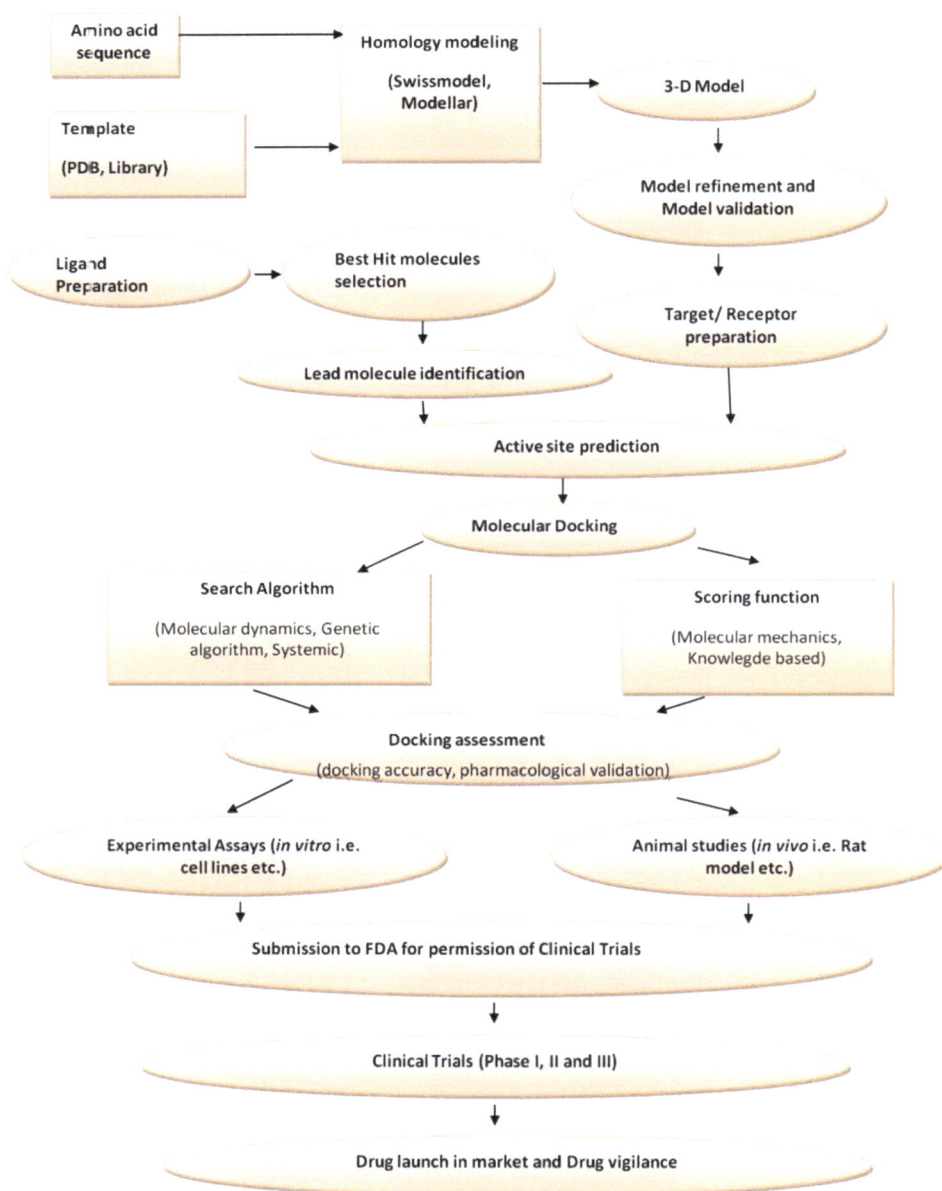

Fig. (1). Flow chart depicting various steps involved in the drug discovery process converting hits to successful leads.

Molecular modelling is an imperative tool when the structure of the target macromolecule especially protein(s) is unknown. It is also termed as comparative modeling, applied to construct the 3D-structure of the protein sequence using freely available online tool such as SWISS-MODEL Workspace (Fig. **2)** [3]. It is important for medicinal chemists to understand the structure-activity relationship. There are different packages available in the market for modelling such as Autodock suite [4], Schrodinger [5], and Patch dock [6].

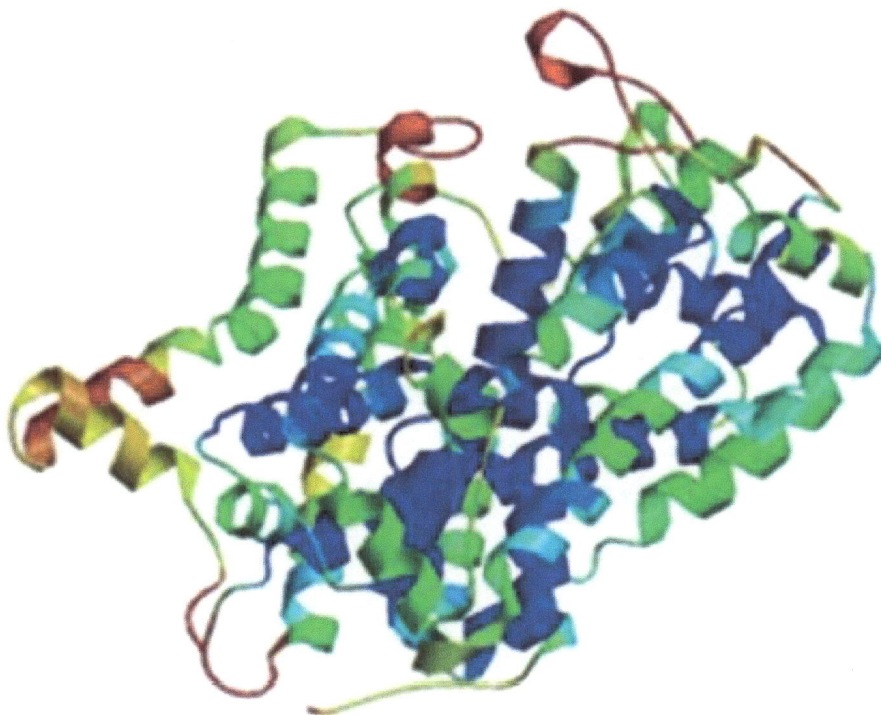

Fig. (2). Generation of the three-dimensional model of sodium-dependent serotonin transporter of *Homo sapiens*, responsible for neurological disorder using SWISS-MODEL.
(Source: Bansal H, Jabalia N. Asian J Pharm Clin Res. 2017;10(8): 299-303)

Docking is another effective tool where the ligands/inhibitors interact with the three-dimensional structure of the target macromolecules. It can be either blind docking when the binding site in the macromolecule is not known or it can be specific docking where the grid is generated around the active site residues [7]. This is an effective strategy to score the binding energies of different compounds. A comparison of the Molecular Mechanics-Generalized Born Surface Area (MM-GBSA) score clearly indicates the binding affinity of the compounds. MM-GBSA score takes in to consideration the free binding energies and enthalpies of macromolecule-ligand interactions [8]. Recently, we used the docking and simulation approaches (Fig. **3**) effectively to identify novel inhibitors (both from natural and synthetic sources) against an important human cell cycle regulating kinase known as microtubule-associated serine/threonine kinase (MASTL) [9]. Thus, docking helps to identify potential hits that can be further optimized.

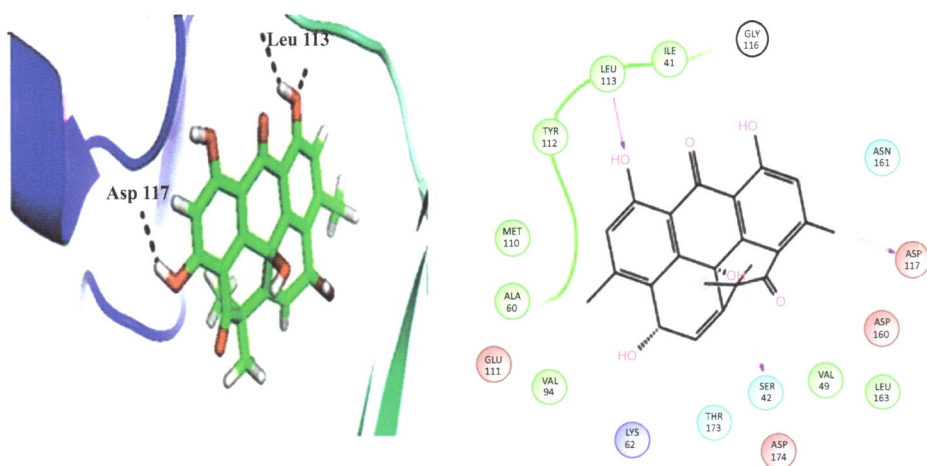

Fig. (3). Docking pose of the best natural compound Resistofavin (UNPD72628) in complex with human MASTL protein.
(Source: Ammarah U, Kumar A, Pal R, Bal NC, Misra G. Scientific reports. 2018;8(1):4894).

Combining all these computational approaches has proved to be effective in saving time in pre clinical trials, decreasing the numbers of animals required for testing at initial stages of the drug discovery process, handling of huge data, and improving the accuracy. Thus, *in silico* methods are cost-effective strategies for drug design [10].

PRINCIPLE

Modeling is used for the prediction of the three-dimensional structure of a given protein sequence referred to as the target based primarily on its alignment to one or more known protein structures that are named as templates. When the template sequences are available for structure modelling, it is known as homology modelling. However, if there are no known structures available for a target sequence then the *ab initio* threading approaches are used for modelling. This is an iterative process that takes in-to consideration fold assignment, target-template alignment, model building, and model evaluation. Recent times have seen an upsurge in the modelled structures with an increase in the number of known protein structures available as templates, resulting from advancement in experimental methods and modeling software. Still, modelling the structures with poor sequence similarities, loops and side chains, rigid body shifts and distortions remains a challenge that needs continuous efforts for solution. Also, it is important to integrate the information from genome sequencing projects, functional and structural genomics for the development of robust, automated, sensitive and complete sequence coverage modelling methods [11].

The classical computational approaches used in drug designing are briefly outlined below:

Homology Modelling

As mentioned above, presence of homologous structures or sequence similarity is must for this technique. This similarity should be approximately above 40%. The commonly used software are Modeller [12], SWISS-model [13] *etc.*. Table **1** enlists the numerous molecular modeling programs based on different approaches used for generating 3-D model.

Table 1. Various modeling tools and programs.

S.No.	Program/Tool	Links
1	Chimera	http://www.marcsaric.de/index.php/Chimera
2	SwissModel	https://swissmodel.expasy.org/
3	Pymol	https://pymol.org/2/
4	Modellar	https://salilab.org/modeller/
5	ModWeb	https://modbase.compbio.ucsf.edu/modweb/
6	VMD	http://www.ks.uiuc.edu/Research/vmd/
7	Biodesigner	http://www.pirx.com/biodesigner/index.shtml
8	XTALVIEW	http://www.sdsc.edu/CCMS/Packages/XTALVIEW/xtalview.html

Molecular Docking

This involves mapping the interactions between a ligand and a target molecule (protein/DNA). Several conformations are generated, amongst which the lowest energy conformations are chosen for analysis [14, 15]. The software routinely used for this purpose are ArgusDock, DOCK, FRED, eHITS, AutoDock and FTDock [16]. The various docking programs and tools are enlisted in Table **2**.

Virtual High-Throughput Screening

This is a high throughput *in silico* method that involves screening of large compound libraries/ databases to obtain the most potential hits that bind to specific target molecules [17]. However, it is more economical and faster as compared to experimental high throughput approaches [18]. Both approaches can be used in parallel to obtain potent leads that are specific for target molecules. The three-dimensional pharmacophore mapping is also generated sometimes at the initial stages to be used for virtual screening later.

Table 2. Summary of docking tools and programs.

S.No.	Docking tools/ Programs	Links
1	DOCK	http://dock.compbio.ucsf.edu/
2	AutoDock	http://autodock.scripps.edu/
3	FelxX	https://www.biosolveit.de/download/
4	Surflex	https://omictools.com/surflex-dock-tool
5	GOLD	https://www.ccdc.cam.ac.uk/solutions/csd-discovery/components/gold/
6	ICM	https://www.molsoft.com/download.html
7	Glide	https://www.schrodinger.com/glide
8	Autodock Vina	http://vina.scripps.edu/
9	McDock	https://github.com/andersx/mcdock
10	UCSFDock	http://dock.compbio.ucsf.edu/
11	FRED	https://omictools.com/fred-tool
12	MOE-Dock	https://www.chemcomp.com/MOE-Structure_Based_Design.htm
13	LeDock	http://www.lephar.com/software.htm
14	rDock	http://rdock.sourceforge.net/
15	HEX	http://hex.loria.fr/
16	ZDock	http://zdock.umassmed.edu/software/
17	MEGADOCK	http://www.bi.cs.titech.ac.jp/megadock/
18	GRAMMX	http://vakser.compbio.ku.edu/resources/gramm/grammx
19	BiGGER	https://sites.fct.unl.pt/biologicalchemistryatfctunl/pages/chemera-bigger
20	GEMDOCK	http://gemdock.life.nctu.edu.tw/dock/download.php
21	PatchDock	https://bioinfo3d.cs.tau.ac.il/PatchDock/

Quantitative Structure-Activity Relationship (QSAR)

It is used to establish a quantitative relationship between structural and physical-chemical properties of the compounds such as topology, electronic and hydrophobic nature [19]. The two techniques used for 3D QSAR are Comparative molecular field analysis (CoMFA) [19, 20] and Comparative Molecular Similarity Indices Analysis (CoMSIA) are recognized as ones of the novel 3D-QSAR approaches. The former deals only with steric and electrostatic properties whereas the latter also includes the hydrogen bonding potentials [21].

Conformational Analysis

The inherent flexibility in most of the compounds is responsible for the possibility of the generation of several conformational isomers differing in energy. Mostly, the conformer with the global minimum energy is chosen for docking purposes. However, it is essential to mention that it does not need to be necessary that the conformation with minimum global energy will be the biologically active compound. Therefore, conformational analysis is done using several approaches such as Utthu *etc.* Each of these methods again includes various approaches like random searches, Monte-Carlo methods, neural networks, distance geometry, genetic and evolutionary algorithms *etc.* to identify the biologically active minimum energy 3D pharmacophore that can be used for docking purposes [22].

Monte-Carlo Simulation

It applies the statistical mechanics as basic principles that result in the prediction of different conformations possessing appropriate thermodynamic, structural and numerical properties which are the weighted average of all these properties present in various conformers. The method is useful in calculating binding free energies of all the poses across the protein surface thus resulting in the identification of all the possible ligand binding sites present in the protein. The binding affinity of the various ligands against a protein target can be ranked with this method [23].

Molecular Dynamic (MD) Simulations

It takes in to consideration the molecular motion at the atomistic level by integrating Newton's equations of motion for each atom. This is achieved by gradually increasing the position, and speed of each atom with a small increment in time, helping in sampling the configuration space. The movement in all the atoms, ions, and solvent molecules is evaluated. This approach is useful for filtering out the artifacts binding sites from the hotspots, calculation of binding free energy, root mean square deviations of different conformations with respect to time and temperature, *etc.* [24]. It is used as a final step for the validation of top-ranking compounds generated after docking [9]. Also, protein-protein/nucleic acid interactions are an important biological phenomenon [25]. Replica exchange molecular dynamics and free energy calculations based on the adaptive biasing force (ABF) method are used for calculating the free energies of interactions when polysorbate 20 interacts with N-phenyl-1-naphthylamine dye. These free energies provide insight in to the morphology of micellar formation by polysorbate 20 [26]. Similar methodology is also used to determine the free energy changes occurring when MEEVD motif present in the C-terminal peptide from the heat shock protein 90 (Hsp90) interacts with the tetratricopeptide repeat A (TPR2A) domain of the heat shock organizing protein (Hop), revealing the details of their binding and unbinding mechanism [27].

Generation of Chemical Structures

It is important to generate chemical structures of compounds that can be potentially screened for their binding affinity towards target proteins. An open source drug discovery platform (http://crdd.osdd.net/optim.php) provides a single point interface for various software used for this purpose [28].

Molecular Structure Visualization

Any computational tool applied for the drug discovery approach is incomplete without structural visualization of results. The structures or docked poses can be represented in the form of space-filling models, surface displays, ray diagrams, or ball and stick models. Some of the common platforms used for this purpose are pymol [29] and Rasmol [30].

Determination of Molecular Properties

It is imperative to determine the molecular properties (physical, chemical and biological) of different chemical compounds including pharmaceuticals for drug development. Molecular properties are essential indicators of many chemical compounds in the pharma industry. We listed the major computational methods used for the calculation of molecular properties in Table **3**.

ADMET Properties

ADMET properties of the identified hits are studied for evaluating their drug likeliness [9]. The various software available for the calculation are included in http://crdd.osdd.net/admet.php [28].

APPLICATIONS OF MOLECULAR MODELING

Molecular modeling has become a valuable and essential tool for medicinal chemists in the drug design process. Molecular modeling describes the generation, manipulation or representation of three-dimensional structures of molecules and associated physico-chemical properties.

Determination of Mechanism of Action of Drugs

Drug resistance to existing therapies against various diseases is a major bottleneck in the drug discovery process. *In silico* approaches are very helpful in gaining information related to resistance mechanisms. For instance, molecular simulations (MD) were recently used to understand the mechanism of resistance to aminoglycosidic antibiotics. The resistance was attributed to the mutations in the bacterial ribosomal A-site [2]. The same lab also studied the effect of changes in the ribosome leading to the resistance to antibiotic telithromycin using a combined Grand Canonical Monte Carlo (GCMC)/Molecular Dynamics (MD) simulation methodology. The insights gained from these studies were effective for improvising the activity of new macrolide analogs with the potential to decrease the resistance.

Table 3. Computational methods for the determination of molecular properties.

S.No.	Methods	Description
1.	*Empirical (molecular mechanics)*	Molecular Mechanics (MM) force fields are the methods of choice for protein simulations, which are essential in the study of conformational flexibility, which approximate the quantum mechanical energy surface with a classical mechanical model, thereby decreasing the computational cost of simulations on the large system by orders of magnitude.
2.	*Molecular dynamics*	Molecular dynamics can be used to explore patterns, strength, and properties of protein behavior, drug–receptor interactions, the solvation of molecules along with conformational rearrangements under various conditions of molecules and their interactions with other molecular species in a range of environments.
3.	*Quantum mechanics*	Quantum mechanics (QM) methods are playing a vital role in computer aided drug design and development because they are used to study molecular systems pertinent to biology, focusing on protein–ligand docking, protein–ligand binding affinities and ligand binding efficacy.

In our lab, we could successfully establish the mechanism of action of platinum based anticancer drugs commonly used against various squamous cell carcinomas. The modelling and simulations based studies effectively established that the drugs bind to DNA in the major groove resulting in a binary complex with conformational changes in DNA that facilitates further binding to the high-mobility group box 1 (HMGB1) proteins [31] (Fig. **4**).

Identification of Novel Drug Targets

The rising resistance against various pathogens and diseases necessitates the identification of novel drug targets. There are several recent studies where *in silico* genome and pathway analysis has resulted in the identification of novel drug targets, for *e.g.* in *Leptospira interrogans* [32], *Pseudomonas aeruginosa* [33], fish pathogen *Edwardsiella tarda* [34], and *Plasmodium falciparumetc*. In *P. falciparum*, the research group implied homology modeling and molecular dynamic simulations on one of the potential drug targets, aminodeoxychorismate lyase, predicting its three-dimensional structure and further virtually screening for its potential inhibitors [35]. On similar lines, another research group focused on the identification of *new* antibiotic targets to circumvent the resistance issue used bioinformatics approaches to screen various databases resulting in the identification of seven enzymes involved in bacterial metabolic pathways and 15 non-homologous proteins membranes proteins in the Gram-positive bacterium *Staphylococcus aureus* [36]. Another study identified heme oxygenase as a novel bacterial antibiotic target. CADD techniques were used for the identification of inhibitors of the bacterial heme oxygenase from *P. aeruginosa* and *Neisseria meningitidis*, thus validating heme oxygenase as a novel antimicrobial target [2].

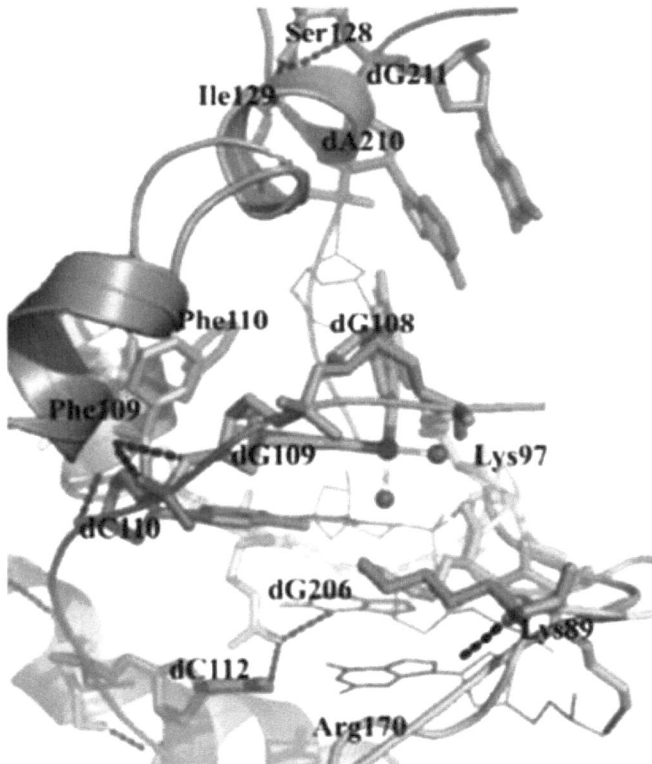

Fig. 4. The binary complex of a high-mobility group of protein domain B1 (HMGB1) with DNA.
(Source: Misra G, Gupta S, Jabalia N. Interdisciplinary Sciences: Computational Life Sciences. 2016 30:1-0).

Determination of Biological Activities of Unknown Compounds

The modelling approaches find wide usage in the design, modification and discovery of new chemical compounds. A comparative understanding of binding energies predicted using various docking tools is helpful in the determination of biological activities of unknown compounds in comparison to the known drugs. In one of the studies various oxadiazole derivatives were characterized for their physical and chemical properties for their use as antibacterial, anti- *Trypanosoma cruzi*, and antifungal agents applying docking approaches [37].

Design of Novel Drugs

Computer-aided drug design approaches are effectively used for the design of novel drugs. Recent SAR studies on one of the G protein-coupled receptors (GPCRs) namely MRGPRX2 resulted in the identification of potential peptides and small molecules as agonists that exhibit selective efficacy in the nanomolar range. This opioid receptor plays an important function in the degranulation of the mast cells [38]. A virtual screening strategy is also used for the designing of novel myocilin inhibitors with implications for glaucoma treatment [39]. Computational approaches have been very effective in establishing the drug likeliness of virtually screened molecules such as proton pump inhibitors [40].

ADVANCED TECHNIQUES

The current advances in computer aided drug design research are discussed below:

Multi-Target Drugs Discovery

Multi-target drug discovery has recently gained momentum for addressing drug resistance issues. The US Food and Drug Administration has recently (September 2017)approved new molecular entities that contain 21% multi-target drugs [41]. Drugs for mood disorder, aspirin, *etc.* are the salient examples under this category. This approach proved to be significant in Alzheimer's disease [42] where multifactorial parameters are involved. Drug repositioning is implied to enhance the effect of existing drugs against various targets [43]. Various other approaches used in multi-target drug discovery include cheminformatics, virtual screening, pharmacophore mapping, molecular docking and molecular dynamics [41].

Site-Identification by Ligand Competitive Saturation

Pharmacophore modelling is an additional approach besides docking used for virtual screening. Recent advances include site identification by ligand competitive saturation (SILCS). This has an advantage as compared to the classical techniques that it takes in to consideration the protein flexibility and desolvation. Besides water, other probe molecules such as benzene, propane, methanol, formamide, *etc.* are used as hydrogen bond donors and acceptors that mutually compete for the active site present in the target protein. This results in the generation of FragMaps reflecting the binding patterns of the probe molecules on the target proteins. Further, Boltzmann transformation is applied to generate grid free energy (GFE) FragMaps for quantitative applications. Additional resources are included in references for further reading that provides a snapshot of computational methods based on biological networks that are helpful in *in silico* drug designing process [44 - 47].

CONCLUDING REMARKS

Drug discovery and development is a long, expensive and multifarious approaches requiring process. Computational methods at the preliminary stages used for screening as well as at later points for validation have proved to be an effective strategy in streamlining this process. This has been exploited on a large scale in medicinal chemistry, pharma industry, biological sciences, and clinical studies. The main advantages are cost-effectiveness, and saving time and resources. There is an upsurge in the open source platforms that provide one point access to users for conducting different *in silico* experiments. It is advised to use these approaches at an earlier stage to avoid late stage failure of drugs that have reached the clinical trials.

CONSENT FOR PUBLICATION

Not applicable.

CONFLICT OF INTEREST

The author confirms that this chapter has no conflict of interest.

ACKNOWLEDGEMENTS

Declared none.

REFERENCES

[1]   Ooms F. Molecular modeling and computer aided drug design. Examples of their applications in medicinal chemistry. Curr Med Chem 2000; 7(2): 141-58.
[http://dx.doi.org/10.2174/0929867003375317] [PMID: 10637360]

[2]   Yu W, MacKerell AD. Computer-Aided Drug Design Methods. New York, NY: In Antibiotics Humana Press 2017; pp. 85-106.

[3]   Bansal H, Jabalia N. In silico characterization and molecular modeling of sodium dependent serotonin

transporter protein from Homo sapiens. Asian J Pharm Clin Res 2017; 10: 299-303.
[http://dx.doi.org/10.22159/ajpcr.2017.v10i8.18954]

[4]   Forli S, Huey R, Pique ME, Sanner MF, Goodsell DS, Olson AJ. Computational protein-ligand docking and virtual drug screening with the AutoDock suite. Nat Protoc 2016; 11(5): 905-19.
[http://dx.doi.org/10.1038/nprot.2016.051] [PMID: 27077332]

[5]   Friesner RA, Murphy RB, Repasky MP, *et al.* Extra precision glide: docking and scoring incorporating a model of hydrophobic enclosure for protein-ligand complexes. J Med Chem 2006; 49(21): 6177-96.
[http://dx.doi.org/10.1021/jm051256o] [PMID: 17034125]

[6]   Schneidman-Duhovny D, Inbar Y, Nussinov R, Wolfson HJ. PatchDock and SymmDock: servers for rigid and symmetric docking. Nucleic Acids Res 2005; 33(Web Server issue)W363-7
[http://dx.doi.org/10.1093/nar/gki481] [PMID: 15980490]

[7]   Gupta S, Misra G, Pant MC, Seth PK. Prediction of a new surface binding pocket and evaluation of inhibitors against huntingtin interacting protein 14: an insight using docking studies. J Mol Model 2011; 17(12): 3047-56.
[http://dx.doi.org/10.1007/s00894-011-1005-8] [PMID: 21360185]

[8]   Zhang X, Perez-Sanchez H, Lightstone FC. A Comprehensive Docking and MM/GBSA Rescoring Study of Ligand Recognition upon Binding Antithrombin. Curr Top Med Chem 2017; 17(14): 1631-9.
[http://dx.doi.org/10.2174/1568026616666161117112604] [PMID: 27852201]

[9]   Ammarah U, Kumar A, Pal R, Bal NC, Misra G. Identification of new inhibitors against human Great wall kinase using *in silico* approaches. Sci Rep 2018; 8(1): 4894.
[http://dx.doi.org/10.1038/s41598-018-23246-0] [PMID: 29559668]

[10]   Wasko MJ, Pellegrene KA, Madura JD, Surratt CK. A role for fragment-based drug design in developing novel lead compounds for central nervous system targets. Front Neurol 2015; 6: 197.
[http://dx.doi.org/10.3389/fneur.2015.00197] [PMID: 26441817]

[11]   Kimko H, Pinheiro J. Model-based clinical drug development in the past, present and future: a commentary. Br J Clin Pharmacol 2015; 79(1): 108-16.
[http://dx.doi.org/10.1111/bcp.12341] [PMID: 24527997]

[12]   Eswar N, Webb B, Marti-Renom MA, *et al.* Comparative protein structure modeling using Modeller. Curr Protoc Bioinformatics 2006; Chapter 5: 6.
[PMID: 18428767]

[13]   Waterhouse A, Bertoni M, Bienert S, *et al.* SWISS-MODEL: homology modelling of protein structures and complexes. Nucleic Acids Res 2018; 46(W1): W296-303.
[http://dx.doi.org/10.1093/nar/gky427] [PMID: 29788355]

[14]   Lesk AJM. Introduction to bioinformatics. 2002.

[15]   Pedro F, Lei H. A systematic review on computer-aided drug design: docking and scoring. J Macao Politech Inst 2010; 4: 47-51.

[16]   Gabb HA, Jackson RM, Sternberg MJ. Modelling protein docking using shape complementarity, electrostatics and biochemical information. J Mol Biol 1997; 272(1): 106-20.
[http://dx.doi.org/10.1006/jmbi.1997.1203] [PMID: 9299341]

[17]   Kövesdi I, Dominguez-Rodriguez MF, Orfi L, *et al.* Application of neural networks in structure-activity relationships. Med Res Rev 1999; 19(3): 249-69.
[http://dx.doi.org/10.1002/(SICI)1098-1128(199905)19:3<249::AID-MED4>3.0.CO;2-0]   [PMID: 10232652]

[18]   Suh M, Park S, Jee H. Comparison of QSAR Methods (CoMFA, CoMSIA, HQSAR) of Anticancer 1-N-Substituted Imidazoquinoline-4,9-dione Derivatives. Bull Korean Chem Soc 2002; 23: 417-22.
[http://dx.doi.org/10.5012/bkcs.2002.23.3.417]

[19]   Wold S, Ruhe A, Wold H, *et al.* The collinearity problem in linear regression. The partial least squares

approach to generalized inverse. SIAM J Sci Statist Comput 1984; 5: 735-43.
[http://cx.doi.org/10.1137/0905052]

[20] Malinowski ER, Howery DG. Factor Analysis in Chemistry. New York: Wiley 1988.

[21] Allen MP, Tildesley DJ. Computer Simulation of Liquids. Oxford, UK: Oxford Science Publications 1989; p 385.

[22] Uthuppan J, Soni K. Conformational analysis: a review. Int J Pharm Sci Res 2013; 4: 34.

[23] Clark M, Guarnieri F, Shkurko I, Wiseman J. Grand canonical Monte Carlo simulation of ligand-protein binding. J Chem Inf Model 2006; 46(1): 231-42.
[http://dx.doi.org/10.1021/ci050268f] [PMID: 16426059]

[24] Hernández-Rodríguez M, Rosales-Hernández MC, Mendieta-Wejebe JE, Martínez-Archundia M, Basurto JC. Current tools and methods in molecular dynamics (MD) simulations for drug design. Curr Med Chem 2016; 23(34): 3909-24.
[http://dx.doi.org/10.2174/0929867323666160530144742] [PMID: 27237821]

[25] Microscopic Perspectives on Macromolecular Interactions: Proteins and Nucleic Acids, Reference Module in Chemistry, Molecular Sciences and Chemical Engineering. Elsevier 2018.

[26] Lapelosa M, Patapoff TW, Zarraga IE. Molecular simulations of micellar aggregation of polysorbate 20 ester fractions and their interaction with N-phenyl-1-naphthylamine dye. Biophys Chem 2016; 213: 17-24.
[http://dx.doi.org/10.1016/j.bpc.2016.03.003] [PMID: 27085143]

[27] Lapelosa M. Free energy of binding and mechanism of interaction for the MEEVD-TPR2A peptide–protein complex. J Chem Theory Comput 2017; 13(9): 4514-23.
[http://dx.doi.org/10.1021/acs.jctc.7b00105] [PMID: 28723223]

[28] Årdal C. Røttingen JA. Open source drug discovery in practice: a case study. PLoS Negl Trop Dis 2012; 6(9)e1827
[http://dx.doi.org/10.1371/journal.pntd.0001827] [PMID: 23029588]

[29] DeLano WL. The PyMOL molecular graphics system http://www. pymol. org2002.

[30] Sayle RA, Milner-White EJ. RASMOL: biomolecular graphics for all. Trends Biochem Sci 1995; 20(9): 374-6.
[http://dx.doi.org/10.1016/S0968-0004(00)89080-5] [PMID: 7482707]

[31] Misra G, Gupta S, Jabalia N. Understanding the Interactions of High-Mobility Group of Protein Domain B1 with DNA Adducts Generated by Platinum Anticancer Molecules Using In Silico Approaches. Interdiscip Sci 2016; 1-0.
[PMID: 27900730]

[32] Amineni U, Pradhan D, Marisetty H. In silico identification of common putative drug targets in Leptospira interrogans. J Chem Biol 2010; 3(4): 165-73.
[http://dx.doi.org/10.1007/s12154-010-0039-1] [PMID: 21572503]

[33] Perumal D, Lim CS, Sakharkar MK. In silico identification of putative drug targets in pseudomonas aeruginosa through metabolic pathway analysis.
[http://dx.doi.org/10.1007/978-3-540-75286-8_31]

[34] Neema M, Karunasagar I, Karunasagar I. In silico identification and characterization of novel drug targets and outer membrane proteins in the fish pathogen Edwardsiella tarda. Open Access Bioinformatics 2011; 3: 37-42.

[35] Rout S, Fatra NP, Mahapatra RK. An in silico strategy for identification of novel drug targets against Plasmodium falciparum. Parasitol Res 2017; 116(9): 2539-59.
[http://dx.doi.org/10.1007/s00436-017-5563-2] [PMID: 28755265]

[36] Hossain M, Chowdhury DUS, Farhana J, *et al.* Identification of potential targets in Staphylococcus aureus N315 using computer aided protein data analysis. Bioinformation 2013; 9(4): 187-92.

[http://dx.doi.org/10.6026/97320630009187] [PMID: 23519164]

[37]   Bassyouni F, El Hefnawi M, El Rashed A, *et al.* Molecular Modeling and Biological Activities of New Potent Antimicrobial, Anti-Inflammatory and Anti-Nociceptive of 5-Nitro Indoline-2-One Derivatives. Drug Des 2017; 6: 2169-0138.
       [http://dx.doi.org/10.4172/2169-0138.1000148]

[38]   Lansu K, Karpiak J, Liu J, *et al.* In silico design of novel probes for the atypical opioid receptor MRGPRX2. Nat Chem Biol 2017; 13(5): 529-36.
       [http://dx.doi.org/10.1038/nchembio.2334] [PMID: 28288109]

[39]   Tang M, Fu Y, Fan Y, *et al.* In-silico design of novel myocilin inhibitors for glaucoma therapy. Trop J Pharm Res 2017; 16: 2527-33.

[40]   Li X, Kang H, Liu W, Singhal S, *et al.* In silico design of novel proton-pump inhibitors with reduced adverse effects. Front Med 2018; 1-8.
       [PMID: 29845582]

[41]   Ramsay RR, Popovic-Nikolic MR, Nikolic K, Uliassi E, Bolognesi ML. A perspective on multi-target drug discovery and design for complex diseases. Clin Transl Med 2018; 7(1): 3.
       [http://dx.doi.org/10.1186/s40169-017-0181-2] [PMID: 29340951]

[42]   Ambure P, Roy K. CADD modeling of multi-target drugs against Alzheimer's disease. Curr Drug Targets 2017; 18(5): 522-33.
       [http://dx.doi.org/10.2174/1389450116666150907104855] [PMID: 26343117]

[43]   Talevi A. Multi-target pharmacology: possibilities and limitations of the "skeleton key approach" from a medicinal chemist perspective. Front Pharmacol 2015; 6: 205.
       [http://dx.doi.org/10.3389/fphar.2015.00205] [PMID: 26441661]

[44]   Wu Z, Li W, Liu G, Tang Y. Network-Based Methods for Prediction of Drug-Target Interactions. Front Pharmacol 2018; 9: 1134.
       [http://dx.doi.org/10.3389/fphar.2018.01134] [PMID: 30356768]

[45]   Li P, Fu Y, Wang Y. Network based approach to drug discovery: a mini review. Mini Rev Med Chem 2015; 15(8): 687-95.
       [http://dx.doi.org/10.2174/1389557515666150219143933] [PMID: 25694073]

[46]   Vitali F, Cohen LD, Demartini A, *et al.* A network-based data integration approach to support drug repurposing and multi-target therapies in triple negative breast cancer. PLoS One 2016; 11(9)e0162407
       [http://dx.doi.org/10.1371/journal.pone.0162407] [PMID: 27632168]

[47]   Leung EL, Cao ZW, Jiang ZH, Zhou H, Liu L. Network-based drug discovery by integrating systems biology and computational technologies. Brief Bioinform 2013; 14(4): 491-505.
       [http://dx.doi.org/10.1093/bib/bbs043] [PMID: 22877768]

CHAPTER 7

Biomolecular Crystallography and Its Applications

Nagendra Singh[*]

School of Biotechnology, Gautam Buddha University, Greater Noida, UP-201312, India

Abstract: X-ray crystallography has immensely contributed to the growth of the science of understanding the three-dimensional structure of matters. The atomic arrangement of small molecules such as salts, inorganic, organic complexes, and metallic compounds was determined. Later on, one after another flood of molecular structures from biological origins was solved using X-ray crystallography. The structure of DNA was determined using fiber diffraction methods in the 1950s, subsequently, structures of polysaccharides, fibrous proteins, and virion particles were determined. The crystal structures of the first protein molecules in the form of lysozyme, myoglobin, and hemoglobin were the enormous achievements of the 1960s, solved by single-crystal diffraction methods. Within a couple of decades later, atomic structures of viruses and membrane receptors were started to be determined. Currently, there are over 125 thousand crystal structures submitted to the PDB database at the rate of more than 3 thousand structures per year. In contrast, there are 12 thousand structures solved by NMR spectroscopy at the rate of just over 100 structures per year, whereas there are only 2 thousand structures available in PDB which are solved using computational methods. It shows the popularity of X-ray crystallography for revealing the atomic details of protein molecules in the field of structural biology. For determining the structure, the molecule is first crystallized to have a repetitive and regular arrangement of arrays in three-dimensional space. As the X-rays have a wavelength in the order of bond distances existing in matters, they are the suitable electromagnetic radiations to be used for finding detailed atomic positions. A beam of X-rays is diffracted from the crystalline matter and is collected at certain positions. The intensities, amplitude, and phases of the diffracted X-rays are convoluted to calculate the electron density of atoms in the crystal. The atomic positions are refined by putting them at mean positions in the electron density and eventually the atomic coordinates in 3-D space are revealed, which define the shape of the matter or a molecule. Biomolecular crystallography deals with the crystal structure determination of biomolecules such as proteins, nucleic acids, polysaccharides, complexes, *etc.* As the structure and the function of a biomolecule are closely associated, revealing the structure is incredibly advantageous in order to understand or alter the function of the biomolecules. This understanding has given rise to the advent of structure-based drug discovery methods. The available 3-D structure of a druggable target protein may also be used for structure-based drug design against a pathophysiological state.

[*] **Corresponding author Nagendra Singh:** School of Biotechnology, Gautam Buddha University, Greater Noida, UP-201312, India; Tel/Fax: 91-120-2344277; E-mails: nagendra@gbu.ac.in, nagendratomar@gmail.com

Keywords: Biomolecules, Braggs law, Crystallization, Drug design, Diffraction, Phase problem, Structure factor, Three-dimensional Structure, X-ray scattering.

INTRODUCTION

X-ray crystallography is a very powerful technique for visualizing atomic and molecular details of molecules from crystals. The most important factor in the deciphering shape of an object is the nature of electromagnetic radiation used in visualization. Microscopy techniques are unable to provide atomic details of biomolecules due to limitations of the wavelength of the radiation used. The wavelength of X-rays falls in the range of atomic bond distances, that make it suitable for resolving atomic positions in a molecule. Unfortunately, no mirror can reflect X-rays, therefore having a microscope like an instrument with X-ray is not possible and hence the image building is done indirectly in X-ray crystallography. X-rays are diffracted by planes within crystal according to Bragg's law. The resulting diffracted rays are collected as dark spots on an image detector. This diffracted pattern provides information about the structure of the crystal. The equipment available can measure the only intensity of the spots as the phase information is lost, which is known as the phase problem in crystallography [1 - 7]. As the phases are required for electron density calculations, it has become essential to acquire the phase information. The phase information can be attained directly using Patterson maps for small molecules, or a similar structure can be used for complex molecules, or phases can also be obtained by anomalous scattering methods. The atoms are modeled into an observed electron density map and refined to have the best match. Eventually, all the atomic positions, bond lengths and various other characteristics of the molecule are revealed. Myoglobin, hemoglobin, and lysozymes were the initial proteins whose structures were solved by crystallography methods [8 - 11]. Atomic structural details can also be observed in solution by NMR spectroscopy, which is advantageous for not requiring crystal formation, as crystallization is generally a bottleneck to X-ray crystallography. However, it is very hard to interpret NMR spectra for larger molecules, whereas, there is no size limitation for solving atomic structures using crystallography. Crystal structure of mega Dalton assemblies such as virus particles, membrane pores, and ribosome has been solved by using X-ray crystallography [12 - 15].

PRINCIPLE

The first step in protein crystallography is the production of pure protein, which is used to form crystals. The crystal is exposed to an X-ray beam to collect diffraction data in the form of spots. The integrated intensities of the diffraction spots are used to reconstruct the electron density map within the unit cell in the crystal. The reconstruction is achieved by the Fourier transformation of the diffraction intensities with phase angle assignment. A high degree of completeness, as well as redundancy in diffraction data, is necessary, meaning that all possible reflections are measured multiple times to reduce systematic and statistical errors using an area detector which can collect diffraction data in a large solid angle. The use of high-intensity X-ray sources, such as synchrotron radiation, is an effective way to reduce data collection time and acquire data with greater accuracy. The atoms are placed in the observed electron density and their positions are refined to get the three-dimensional structure of the molecule.

Protein Production

The primary requirement of structure determination is the sample in the purest possible form without micro and macro heterogeneity. In the earlier time, pure proteins used to be obtained by purification from naturally existing sources, which is a very cumbersome and time taking approach and generally employs multiple steps of column chromatography. The advent of recombinant DNA technology methods has simplified protein production to a greater extent. Hence, nowadays, the gene for the protein to be studied is cloned and expressed in a suitable host with a tag sequence. The protein expression is induced in the host and the produced protein is harvested from the host cells, which is purified using single or sometimes additional chromatography techniques depending on the tag present in the protein. The most common tag being used in expression is His tag, which is six histidine amino acids inserted at generally either at N / C or both termini of the protein. Nickel metal immobilized with a linker in the stationary phase of the chromatography in Immobilized Metal Affinity Chromatography (IMAC) is used to purify protein in such cases, as Histidine cluster has an affinity for nickel atoms [1, 6].

Crystallization

Crystal formation is a crucial step in crystallographic studies. There are no known ways to predict the crystallization of any material including biomolecules. The basic principle is to reduce the solvent concentration to aid intermolecular interactions in a solution, which may eventually help in crystal formation. The physical and chemical nature including the flexibility of the molecule affects its crystallization behavior. Small molecules crystallize faster by simply evaporating the solvent whereas, biomolecules such proteins require more complex processes with a large number of hit and trial approaches [1, 16].

A crystal is formed by the periodic assembly of molecules in a three-dimensional array that arises by the separation of solvents from the molecules in the solution. In other words, decreasing solubility of the solvent gives rise to the molecular assembly in a specific, regular and repetitive lattice known as a crystal. Generally, a highly concentrated solution of biomolecule such as 10mg/ml is used and some precipitating agents or combinations of various such reagents are gradually added to reach the super-saturation state of the solution. The solubility of the solute is governed by the solubility curve or phase diagram, which is used to calculate the amount of precipitating agent to be added in a fixed concentration of solute to achieve the super-saturation stage. Till date, it has not been possible to predict the combination of precipitant, ionic strength, pH or temperature which give rise to diffractive crystal formation, hence various permutations and combinations are tried in initial screening to know the precise condition for obtaining crystals. Therefore when any hit is received, the conditions are further optimized in order to get good quality crystals [1, 16, 17]. In a protein crystal, flexible molecules are held together with weak noncovalent forces such as hydrogen bonds, ionic interactions and or Vander Waals interactions, which make protein crystal very fragile and sensitive to their environment. The crystals are then mounted in X-rays and the diffraction data are collected. The diffraction pattern is used to infer about the quality of a crystal.

The vapor diffusion technique is the most popular way of growing protein crystals (Fig. **7.1**). It has two variations known as hanging drop and sitting drop methods. In a hanging drop setup, a drop of few microliters of the protein sample dissolved in appropriate buffer composition mixed with few microliters of the reservoir solution is placed on a glass coverslip. This coverslip is placed in an inverted manner on a well containing the reservoir solution having a precipitating reagent or combinations. The coverslip is sealed airtight using vacuum grease and left to incubate undisturbed at a fixed temperature. As the concentration of reservoir solution is less in the drop in comparison to the well, the solvent evaporates from drop to the reservoir leaving lesser solvent in drop to enhance surface-surface interactions between the protein molecules give rise to crystal formation. Supersaturation is a thermodynamically metastable state, where protein molecules may aggregate specifically to produce crystals or may lead to nonspecific aggregation in the form of precipitation as shown in the phase diagram (Fig. **7.2**). Therefore, all possible variables in protein concentration, buffer composition, precipitant concentration, pH, temperature, *etc.* are tried using multi-well plates to crystallize a protein. In the sitting drop method, the sample drop lies in the concave surface of a microbridge located in the well instead of hanging on to the coverslip [7, 16 - 19].

X-ray Scattering

X-rays are high energy electromagnetic radiations laying between the UV region and gamma rays of the spectrum. Their wavelength lies between 0.4 to 2.5Å

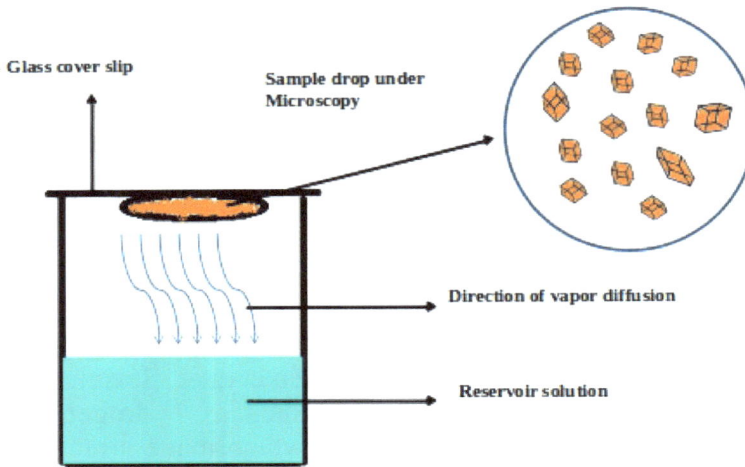

Fig. (7.1). Graphical representation of basic setup used for hanging drop vapor diffusion technique for protein crystallization. The reservoir is generally a well of multi-well plate, filled with cocktail of precipitating reagent. The sample drop is a mixture of protein sample and reservoir solution kept on a siliconized glass cover slip. Due to concentration difference of precipitating agent vapor diffusion occurs from drop to reservoir leading to super saturation state and crystallization of the sample. The crystals generally are micrometer to millimeter size and are seen under a light microscope.

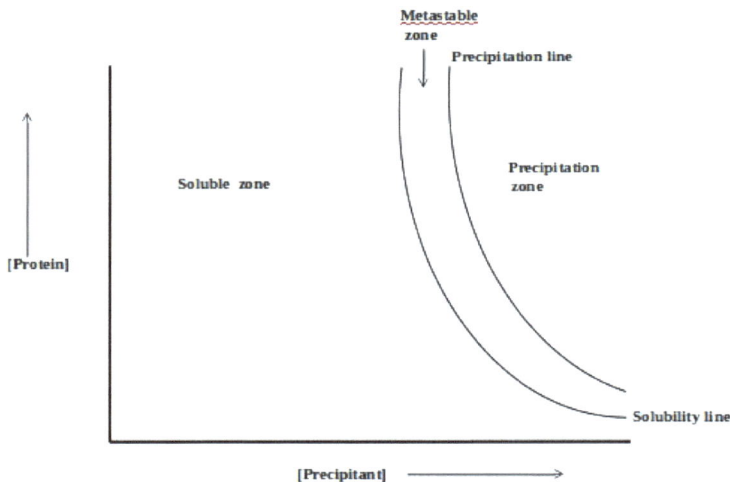

Fig. (7.2). Solubility phase diagram representing effect of precipitation agent on a protein solution. When we keep increasing precipitant concentration in a fixed protein solution, it convert into precipitate. Just before precipitation line, metastable zone or super saturation zone lies, which is also known as nucleation zone. Protein crystallizes by keeping its solution in the metastable zone.

which is of particular interest in the structure determination of biomolecules. As the refractive index of X-rays is close to one, these cannot be focused by means of lenses like ordinary light or electrons and therefore x -rays cannot be used in microscopy to visualize molecular structures.

X-ray scattering from single molecules is too weak to be detected, hence a crystal is required to observe the measurable intensity of the scattered X-rays. A crystal is a periodic arrangement of atoms and molecules where each atom is having multiple exact translation equivalent positions in the space along the direction of the lattice propagation. The smallest repetitive unit of the crystal is called the unit cell. In the case of proteins, a unit cell contains a large number of atoms and electrons [19]. The X-rays diffracted from these electrons interfere with each other, called destructive interference. Whereas, X-ray diffracted from the electron positioned in phase leads to constructive interference which gives rise to intense spot after diffraction. When emitted on the matter, X-rays get absorbed by the electrons to get into high energy transitions. The excited electrons return back to the normal state by emitting X-rays of the same wavelength. These emitted X-rays are collected on an X-ray detector such a photographic film or CCD detectors to record their intensities. The X-ray scattering is dependent on the distribution of electrons in a crystal lattice. The effective number of scattered electrons from a unit cell of the crystal is called structure factor, F. F also depends on the direction of the X-rays scattering. For n number of atoms in the unit cell [7, 19]:

$$F(S) = \sum_{j=1}^{n} f_j \exp(2\pi i r_j . S) \tag{7.1}$$

Where S is a vector perpendicular to the plane reflecting the incident beam at an angle θ, length of $S = 2 \sin \theta / \lambda$. Atom j is at position r_j with respect to the origin of the unit cell. Another unit cell may have its origin at $t \times \mathbf{a}$, $u \times \mathbf{b}$, and $v \times \mathbf{c}$, where t,u and v are whole numbers and a,b and c are cell dimensions. The wave scattered from a crystal is the sum of all scattered waves by all unit cells present in the crystal. Hence, scattering from crystal becomes the product of three summations. The amplitude of wave scattered by the crystal would be [7, 19]:

$$W^{crystal}(S) = F(S) . \sum_{t=0}^{n=1} \exp(2\pi i t \mathbf{a} . S) . \sum_{u=0}^{n=2} \exp(2\pi i u \mathbf{b} . S) . \sum_{v=0}^{n=3} \exp(2\pi i v \mathbf{c} . S) \tag{7.2}$$

In the case of large crystals, the three summations over exponential functions have the property that they are zero unless a.S= h, b.S=k, and c.S=l. Where h,k,l are indices having whole numbers as negative, zero or positive. This is known as 'Lau condition' which needed to be fulfilled for all unit cells to scatter in-phase and the amplitude of the scattering by the crystal is proportional to the amplitude of the scattering factor F. The intensity will be proportional to $|F|^2$. The Lau condition is better defined by "Bragg's law". Bragg diffraction occurs when X-rays are scattered by atoms located in a crystal lattice and undergoes to constructive interference. X-ray scattering occurs by lattice plane separated by distance d in a crystal. The constructive interference or diffraction of the scattered waves occurs only if the difference of path lengths of the waves is equal to an integer multiple of its wavelength, this is known as 'Bragg's Law' for diffraction [1, 7]. It is also stated as:

$$2d \sin \theta = n\lambda \tag{7.3}$$

where θ is the reflecting angle, λ is the wavelength of X-rays and n is an integer. The equation 7.1 represents the wave and $W^{crystal}(S)$ given as the sum of atomic contributions in all unit cells. A more exact expression of the scattered wave can be written as [7, 19]

$$W^{crystal}(S) = \int_{crystal}^{0} \rho(r)\, exp\, (2\pi i r.S)\, dv\ (real) \tag{7.4}$$

where $\rho(r)$ is the electron density distribution in the unit cell and the integration is over all electrons present in all unit cells in the crystal. This is known as Fourier transformation. $W^{crystal}$ is Fourier transform of $\rho(r)$ and the value of $\rho(r)$ can be calculated by inverse Fourier transformations as [7, 19]

$$\rho(r) = \int_{S}^{c} Wcrystal\ (S)\, exp\, (-2\pi i r.S)\, dv\ (reciprocal) \tag{7.5}$$

Considering fractional coordinates x,y,z in the unit cell, integration over the unit cell and the unit cell volume V, the equation can be rewritten as [7, 19]:

$$\rho(xyz) = (1/V)\sum_h \sum_k 1 \sum_l 1 |F(hkl)|\, exp\, [-2\pi i\, (hx+ky+lz) + i\alpha(hkl) \tag{7.6}$$

$\alpha(hkl)$ is the phase angle. Equation 7.5 can be used to calculate electron density of a unit cell when values of $|F(hkl)|$ and $\alpha(hkl)$ are known. Information about phases is lost during diffraction data collection, which is known as a phase problem in crystallography. $|F(hkl)|$ is calculated from the diffraction data whereas the value of $\alpha(hkl)$ cannot be determined directly in case of macromolecules. Therefore, phases are calculated using indirect methods for macromolecular structure determination. There are various methods to calculate phases in macromolecular crystallography such as Molecular Replacement (MR), Multiple Isomorphous Replacement (MIR) and Multiple wavelength Anomalous Dispersion (MAD).

Molecular Replacement

With the increasing number of protein structures in the database, the chances of having some kind of structural similarity with an unknown protein also increase. MR method is employed when a similar structure to the protein for which data collected are available. For example, if the structure of a native protein is available, the structure of the same protein complexes with ligands or mutated forms of the protein is easily solved using MR. In this method, the coordinates of the homologous model are positioned in the unit cell in such a way that the theoretical diffraction from the model resembles maximum with the experimental diffraction pattern from the unit cell of the crystal. The rotational and translational parameters required to have the best agreement in both the structure factors (of the model and the target data) are determined, which are eventually applied to the model coordinates to determine relative phases. Considering X, the set of vectors represent original theoretical structure factors and if X' is the rotated and translated form, then the transformation can be given as [1, 7, 19, 20]:

$$X' = X[R] + T \qquad (7.7)$$

Where R is the matrix representing the rotation of the original positions into new orientation along with the translation vector T. The equation 7.6 can be represented by Fig. (7.3).

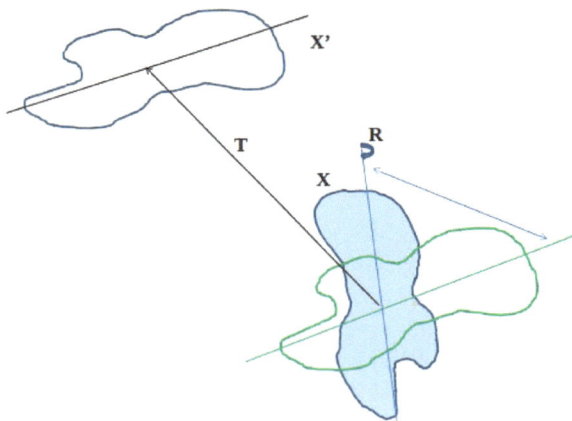

Fig. (7.3). A simple illustration of the equation 7.6 where the X' represents the transformed vector position of the MR model, X is the original model vector, R is rotation matrix, T is translation matrix.

As the number of protein folds for globular proteins is limited in biological systems hence, are expected to have their structures known sooner in the future and then it may be possible to use MR for determining structures of all other protein molecules. MR method will have greater importance in the phase determination of most of the unknown similar molecules in the future. A selected model is expected to be successful only if the sequence identity is generally more than 30%. Higher identity leads to a more successful and accurate phase determination. Along with structural similarity, another most important factor adversely affecting the success of MR is the presence of multi copies of a molecule in the unit cell, which reduces the signal to noise ratio.

APPLICATIONS

X-ray Crystallography in Medicine

X-ray crystallography is an essential and prevailing technique in drug development against diseases, used by pharmaceutical companies in the process of exploring new medicines. A drug is a generally small molecule that recognizes the surface of disease-causing receptor protein and inhibits its function. Medicinal chemists got assistance from the understanding of precise atomic details of small molecules such as vitamins, steroids, folate, *etc.* to modify their structure and to use them for treatment of diseases. Crystallography of drug binding to its receptor allows the study of the specific interactions of the drug and its receptor which in turn provide essential details of its mechanism. Which is further used to design or improve drug molecules for better efficacy and selectivity [19, 21].

The atomic structure of a receptor protein reveals fine details including surface amino acids, stereochemistry, and geometry of active site or binding site located on the protein molecule. If the function of that protein is leading to pathophysiological state, it can form a target for drug development. The essential information from the crystal structure is used to design small ligand compounds with acceptable affinity and selectivity, which are later form drug candidates to treat the disease [22]. Considering this, there is an inconceivable increase in the number of drug design projects in academia and pharmaceutical companies. In fact, major pharmaceutical companies have crystallographic units in these days to assist their drug development programs.

X-ray Crystallography in Genetic Diseases

Crystallography has been proven very useful in deciphering the cause and mechanism of various genetic diseases at atomic and molecular levels. Any structural change in a gene product due to mutations can be studied at the atomic level using crystallography which provides insight about the mechanism of the disease and possibilities of its treatment. Mechanisms of numerous diseases such as sickle cell anemia, thalassemias, amyloidosis, phenylketonuria, *etc.* have been well studied using crystallography. Cystic fibrosis is among frequently occurring diseases in which mutations cause malfunctioning of an ion channel leading to the severity of the pathological state. Crystallography has been very useful in the understanding of the diseases and forming the base for drug development [19, 23].

X-ray Crystallography in Infectious Diseases

Crystallography has solved the atomic structure of several proteins from the bacterial, viral, protozoan origin which could form potential drug targets to treat infectious diseases. There are several such examples from viruses such as HIV, influenza, and hepatitis. The crystal structure of protease, integrase, reverse transcriptase, core and surface proteins from HIV is highly valuable in developing novel therapeutic molecules against the virus [24 - 27]. A structure of influenza neuraminidase and its complexes with ligands has been useful in the understanding of their interactions and further application in designing potential molecules [28].

Similarly, a large number of crystal structures have been solved from infectious bacterial sources in order to understand their atomic details of functioning, which is being exploited for developing antibacterial therapeutic agents. As per the World Health Organization (WHO), 9 million people get infected per year by Mycobacterium tuberculosis, a causative agent of TB. The genomic information of Mtb is available which has made a tremendous impact on the structure determination of a large number of proteins from the bacteria, which are providing insights about the survival of the organism and methods of controlling its growth to treat TB [29]. The structure of bacterial toxins is also being used to design prophylactics to treat the infection.

A major cause of reduced mortality worldwide is an infection from several protozoa such as mainly *Plasmodium* (causing malarial parasite), *Trypanosoma* (causing agent of sleeping sickness and Chagas diseases) and *Leishmania* (causing Kala-azar). Crystal structures of several crucial surface protein, glycolytic enzymes and proteases are available in these protozoa, which are being used for developing small molecules as novel therapeutic agents to treat their infections. Structures of several proteins such as beta-lactamase, HIV protease, HIV reverse transcriptase, DHFR, *etc.* from the drug resistance acquired infectious organisms have given insights about the mechanism of the resistance development and possible ways to tackle it [30 - 34].

X-ray Crystallography in Drug Metabolism

Living organisms use special enzymes to modify and or breakdown foreign substances such as drugs and eventually throw them out of the body, a process known as xenobiotic metabolism. This process of biotransformation occurs in two steps, in the first step the drug is functionalized and in second, it is conjugated with hydrophilic substances in the body to make their excretion easy. This process of detoxification renders several good drug candidates ineffective, hence the knowledge of the structure and functioning of these enzymes involved in drug biotransformation is crucial for drug development. Human Serum Albumin (HSA) has been known to bind drugs in the body to assist the biotransformation process and to lessen the effective drug concentration in the blood. There are more than 100 crystal structures of HSA in native and complex with broad categories of drugs and other compounds available in the protein databank, which have given details of drug interactions with HSA at atomic and molecular levels along with their pharmacokinetics [35, 36].

X-ray Crystallography in Vaccine Development

Vaccines are one of the most effective ways to control the progression of diseases. vaccines have been developed against several diseases. vaccine development can be highly assisted by x-ray crystallography. Three-dimensional crystal structure of viruses is being used to understand the details of the coat proteins and their arrangement which will eventually help in designing vaccines. Loop regions in viral proteins are being identified using x-ray crystallography which can be replaced with antigenic peptides to produce an immune response to aid the vaccine design and development process [37, 38].

ADVANCED TECHNIQUES

The potential future advances in X-ray crystallography will be exciting. The visualization studies on single molecules are not currently possible. Owing to the large complexity of biological molecules, X-ray crystallography and NMR have been the most prominent techniques to understand their three-dimensional structure. Cryoelectron microscopy is another technique that is going through an evolution to be used for molecular details of higher assemblies of biomolecules. High-intensity X-ray sources in the form of synchrotron radiations have permitted the development of dramatically faster and high-quality X-ray data collections. The solution to the phase problem is continued to be a rate-limiting step in X-ray crystallography, which is being overcome by experimental methods combined with high-resolution diffraction data sets. Development in versatile and user-friendly structure determination and molecular visualization programs will be aiding to the speed and accuracy in the structure determination methods.

CONCLUDING REMARKS

X-ray crystallography is based on the interference and diffraction of X-rays arises from their interactions with the electronic clouds of the crystal. The intensity of the diffracted rays is used in the calculation of electron density and eventually the atomic model is built. A summary of the technique is shown in Fig. (**7.4**).

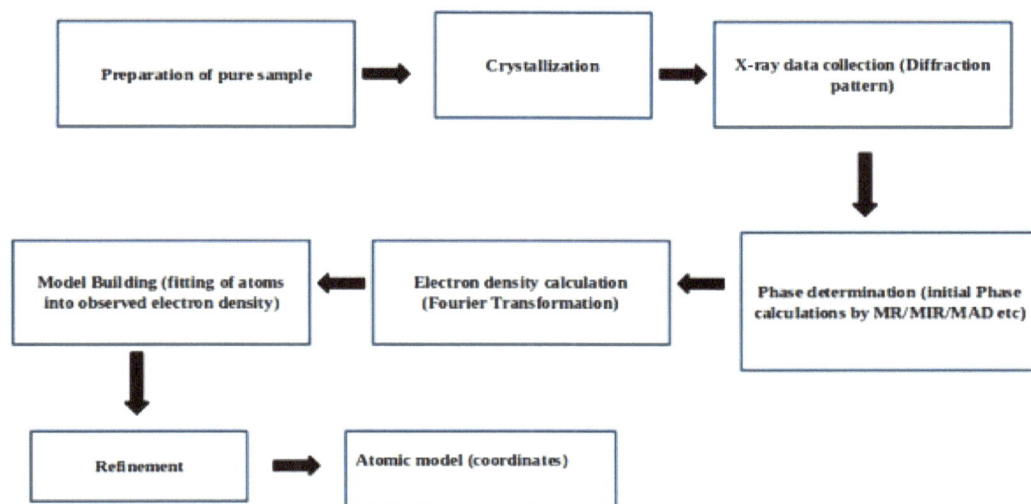

Fig. (7.4). A flowchart summary of the x-ray crystallography technique. The final coordinates are validated based on geometrical and conformational constraints and deposited to the protein data bank (PDB) with a specific accession code.

During initial days, complete knowledge of the techniques from crystallization to the structure determination was necessary in order to solve a biomolecular structure using manual methods. Nowadays, due to the tremendous development in hardware and software used in crystallography, the determination of the crystal structure of any biomolecule has become easier. The need for a crystal and determination of phases have been the major bottlenecks for the crystallography. High throughput robotics is being used for crystallization. Emerging methods for phase determination along with greater accuracy in diffraction data collection using synchrotron radiations have reduced the difficulties for phase determination whereas, getting a good quality crystal still remains a rate-limiting step. Possibilities of collecting data from a single molecule to bypass crystallization may have the greatest impact on the development of the technique.

CONSENT FOR PUBLICATION

Not applicable.

CONFLICT OF INTEREST

The author confirms that this chapter contents have no conflict of interest.

ACKNOWLEDGEMENTS

Declare none.

REFERENCES

[1]   Rupp B. Biomolecular Crystallography. Garland Sciences. LLC: Taylor & Francis Group 2001.

[2]   Authier A. Dynamic theory of X-ray diffraction IUCR Book series. Oxford university press 2001.

[3]   Bragg WH. The analysis of crystal structure by X-rays. Science 1924; 60(1546): 139-49.
        [http://dx.doi.org/10.1126/science.60.1546.139] [PMID: 17750761]

[4]   Bragg WH. X-rays and crystalline structure. Science 1914; 40(1040): 795-802.
        [http://dx.doi.org/10.1126/science.40.1040.795] [PMID: 17828995]

[5]   Bragg WH. The meaning of the crystal. Science 1930; 71(1848): 547-50.
        [http://dx.doi.org/10.1126/science.71.1848.547] [PMID: 17796359]

[6]   Drenth J. Principles of Protein X-ray Crystallography, Springer. Crystallography 1994.
        [http://dx.doi.org/10.1007/978-1-4757-2335-9]

[7]   Giacovazzo C, Ed. Fundamentals of crystallography. 2nd ed., Oxford University Press 2002.

[8]   Strandberg B. Chapter 1: building the ground for the first two protein structures: myoglobin and haemoglobin. J Mol Biol 2009; 392(1): 2-10.
        [http://dx.doi.org/10.1016/j.jmb.2009.05.087] [PMID: 19712775]

[9]   Blake CC, Koenig DF, Mair GA, North AC, Phillips DC, Sarma VR. Structure of hen egg-white lysozyme. A three-dimensional Fourier synthesis at 2 Angstrom resolution. Nature 1965; 206(4986): 757-61.
        [http://dx.doi.org/10.1038/206757a0] [PMID: 5891407]

[10]   Kendrew JC, Bodo G, Dintzis HM, Parrish RG, Wyckoff H, Phillips DC. A three-dimensional model of the myoglobin molecule obtained by x-ray analysis. Nature 1958; 181(4610): 662-6.
        [http://dx.doi.org/10.1038/181662a0] [PMID: 13517261]

[11]   Perutz MF, Rossmann MG, Cullis AF, Muirhead H, Will G, North AC. Structure of haemoglobin: a three-dimensional Fourier synthesis at 5.5-A. resolution, obtained by X-ray analysis. Nature 1960; 185(4711): 416-22.
        [http://dx.doi.org/10.1038/185416a0] [PMID: 18990801]

[12]   Wimberly BT, Brodersen DE, Clemons WM Jr, *et al.* Structure of the 30S ribosomal subunit. Nature 2000; 407(6802): 327-39.
        [http://dx.doi.org/10.1038/35030006] [PMID: 11014182]

[13]   Arnold E, Erickson JW, Fout GS, *et al.* Virion orientation in cubic crystals of the human common cold virus HRV14. J Mol Biol 1984; 177(3): 417-30.
        [http://dx.doi.org/10.1016/0022-2836(84)90293-6] [PMID: 6088778]

[14]   Holmes KC, Stubbs GJ, Mandelkow E, Gallwitz U. Structure of tobacco mosaic virus at 6.7 å resolution. Nature 1975; 254(5497): 192-6.

[http://dx.doi.org/10.1038/254192a0] [PMID: 1113882]

[15]   Narwal M, Singh H, Pratap S, *et al*. Crystal structure of chikungunya virus nsP2 cysteine protease reveals a putative flexible loop blocking its active site. Int J Biol Macromol 2018; 116: 451-62.
[http://dx.doi.org/10.1016/j.ijbiomac.2018.05.007] [PMID: 29730006]

[16]   Bergfors T, Ed. Protein Crystallization; San Diego, CA. USA: International University Line 2009.

[17]   Ducruix A, Giege R, Eds. Crystallization of nucleic acids and proteins; Oxford, UK. USA: Oxford University Press 1999.

[18]   McPherson A. Crystallization of biological macromolecules; Cold Spring HArbor, NY. USA: CSHL Press 1999.

[19]   Rossmann MG, Arnold E, Eds. International tables for crystallography. 1st ed. Kluwer Academic Publishers 2001; Vol. F.

[20]   Rossman MG, Ed. The molecular replacement method. New York: Gordon & Breach 1972.

[21]   Carvalho AL, Trincão J, Romão MJ. X-ray crystallography in drug discovery. Methods Mol Biol 2009; 572: 31-56.
[http://dx.doi.org/10.1007/978-1-60761-244-5_3] [PMID: 20694684]

[22]   Kuntz ID. Structure-based strategies for drug design and discovery. Science 1992; 257(5073): 1078-82.
[http://dx.doi.org/10.1126/science.257.5073.1078] [PMID: 1509259]

[23]   Collins FS. Cystic fibrosis: molecular biology and therapeutic implications. Science 1992; 256(5058): 774-9.
[http://dx.doi.org/10.1126/science.1375392] [PMID: 1375392]

[24]   Boyer PL, Smith SJ, Zhao XZ, *et al*. Developing and Evaluating Inhibitors against the RNase H Active Site of HIV-1 Reverse Transcriptase. J Virol 2018; 13;92(13): pii: e02203-.

[25]   Eckert DM, Malashkevich VN, Hong LH, Carr PA, Kim PS. Inhibiting HIV-1 entry: discovery of D-peptide inhibitors that target the gp41 coiled-coil pocket. Cell 1999; 99(1): 103-15.
[http://dx.doi.org/10.1016/S0092-8674(00)80066-5] [PMID: 10520998]

[26]   Dyda F, Hickman AB, Jenkins TM, Engelman A, Craigie R, Davies DR. Crystal structure of the catalytic domain of HIV-1 integrase: similarity to other polynucleotidyl transferases. Science 1994; 266(5193): 1981-6.
[http://dx.doi.org/10.1126/science.7801124] [PMID: 7801124]

[27]   Erickson J, Neidhart DJ, VanDrie J, *et al*. Design, activity, and 2.8 A crystal structure of a C2 symmetric inhibitor complexed to HIV-1 protease. Science 1990; 249(4968): 527-33.
[http://dx.doi.org/10.1126/science.2200122] [PMID: 2200122]

[28]   Pokorná J, Pachl P, Karlukova E, *et al*. Kinetic, Thermodynamic, and Structural Analysis of Drug Resistance Mutations in Neuraminidase from the 2009 Pandemic Influenza Virus. Viruses 2018; 10(7)E339
[http://dx.doi.org/10.3390/v10070339] [PMID: 29933553]

[29]   Li R, Sirawaraporn R, Chitnumsub P, *et al*. Three-dimensional structure of M. tuberculosis dihydrofolate reductase reveals opportunities for the design of novel tuberculosis drugs. J Mol Biol 2000; 295(2): 307-23.
[http://dx.doi.org/10.1006/jmbi.1999.3328] [PMID: 10623528]

[30]   da Rosa R, de Moraes MH, Zimmermann LA, Schenkel EP, Steindel M, Bernardes LSC. Design and synthesis of a new series of 3,5-disubstituted isoxazoles active against Trypanosoma cruzi and Leishmania amazonensis. Eur J Med Chem 2017; 128: 25-35.
[http://dx.doi.org/10.1016/j.ejmech.2017.01.029] [PMID: 28152426]

[31]   De Gasparo R, Brodbeck-Persch E, Bryson S, *et al*. Biological Evaluation and X-ray Co-crystal Structures of Cyclohexylpyrrolidine Ligands for Trypanothione Reductase, an Enzyme from the

Redox Metabolism of Trypanosoma. ChemMedChem 2018; 13(9): 957-67.
[http://dx.doi.org/10.1002/cmdc.201800067] [PMID: 29624890]

[32]   Lantwin CB, Schlichting I, Kabsch W, Pai EF, Krauth-Siegel RL. The structure of Trypanosoma cruzi trypancthione reductase in the oxidized and NADPH reduced state. Proteins 1994; 18(2): 161-73.
[http://dx.doi.org/10.1002/prot.340180208] [PMID: 8159665]

[33]   Schumacher MA, Carter D, Scott DM, Roos DS, Ullman B, Brennan RG. Crystal structures of Toxoplasma gondii uracil phosphoribosyltransferase reveal the atomic basis of pyrimidine discrimination and prodrug binding. EMBO J 1998; 17(12): 3219-32.
[http://dx.doi.org/10.1093/emboj/17.12.3219] [PMID: 9628859]

[34]   Schumacher MA, Carter D, Roos DS, Ullman B, Brennan RG. Crystal structures of Toxoplasma gondii HGXPRTase reveal the catalytic role of a long flexible loop. Nat Struct Biol 1996; 3(10): 881-7.
[http://dx.doi.org/10.1038/nsb1096-881] [PMID: 8836106]

[35]   Kawai A, Yamasaki K, Enokida T, Miyamoto S, Otagiri M. Crystal structure analysis of human serum albumin complexed with sodium 4-phenylbutyrate. Biochem Biophys Rep 2018; 13: 78-82.
[http://dx.doi.org/10.1016/j.bbrep.2018.01.006] [PMID: 29387812]

[36]   Smolková R, Zeleňák V, Smolko L, *et al.* Novel zinc complexes of a non-steroidal anti-inflammatory drug, niflumic acid: Structural characterization, human-DNA and albumin binding properties. Eur J Med Chem 2018; 153: 131-9.
[http://dx.doi.org/10.1016/j.ejmech.2017.05.009] [PMID: 28502586]

[37]   Kohara M, Abe S, Komatsu T, Tago K, Arita M, Nomoto A. A recombinant virus between the Sabin 1 and Sabin 3 vaccine strains of poliovirus as a possible candidate for a new type 3 poliovirus live vaccine strain. J Virol 1988; 62(8): 2828-35.
[PMID: 2839704]

[38]   Kwong PD, Wyatt R, Robinson J, Sweet RW, Sodroski J, Hendrickson WA. Structure of an HIV gp120 envelope glycoprotein in complex with the CD4 receptor and a neutralizing human antibody. Nature 1998; 393(6686): 648-59.
[http://dx.doi.org/10.1038/31405] [PMID: 9641677]

CHAPTER 8

RNA Sequencing Technology for Biomedical Sciences

Sandeep Ameta[*] and **Roberta Menafra**

Laboratoire de Biochimie, École supérieure de physique et de chimie industrielles de la ville de Paris [ESPCI Paris], France CNRS UMR 8231 Chimie Biologie Innovation, PSL Research University, Paris, France

Abstract: In the last two decades, the development of massive parallel sequencing methods has allowed the sequencing of RNA at an unprecedented resolution, unleashing an enormous wealth of information about the cellular state. Sequencing has accelerated biomedical research by identifying novel mutations, aberrant splicing patterns, splicing isoforms, new gene regulators, and cell-to-cell heterogeneity. In order to efficiently characterize the complexity of the complete transcriptome, there is a steady development for different RNA sequencing [RNA-seq] protocols by improving different steps from library preparation to the data analysis. Furthermore, with the advancements in the sequencing strategies, single-cell RNA sequencing[scRNA-seq] methods have been developed allowing to address the heterogeneity in cell types, and mRNA expression at a remarkable resolution. The majority of these methods involve the conversion of RNA to cDNA and thus amenable to errors, PCR and ligation biases, and inefficiencies of enzymes. Amid these challenges, strategies have been developed to sequence the RNA directly at the single-molecule level which allows to overcome these biases. This chapter provides a brief overview of different sequencing technologies available for the RNA-seq, scRNA-seq and single molecule RNA sequencing along with the different aspects where RNA sequencing has contributed to the biomedical field.

Keywords: Direct RNA sequencing, Different sequencing strategies, Next generation sequencing, RNA-seq, RNA-related diseases, ScRNA-seq.

INTRODUCTION

RNA plays a multitude of roles which are diverse and central to the cellular functions. Owing to the technological advancements in the last decades, our perspective for RNA has changed from being a passive messenger involved in translating the information to one of the critical biomolecules involved in regula-

[*] **Correspondingauthor Sandeep Ameta:** Laboratoire de Biochimie, École supérieure de physique et de chimie industrielles de la ville de Paris (ESPCI Paris), France; Tel: +33 140794587; E-mail: sandeep.ameta@espci.fr

tion [1 - 8], catalysis [4, 9 - 12], metabolism [13, 14], development [15 - 17], diseases [16, 18], and much more. With deeper insights into the cellular processes, it is well established that only a small percentage (1-5%) of transcribed RNA is translated into proteins, the so-called messenger RNA (mRNA), leading to the discovery of new roles for RNAs [19, 20]. Within the cell, various steps are involved in the processing of RNA, and defects in any of these steps can lead to the onset of diseases. One of the key processes is pre-mRNA splicing, where non-coding part of pre-mRNA is excised out by a complex RNA-protein machinery [21, 22]. The sequences in these non-coding regions contain information about exon-exon junction, and interaction with different splicing proteins, thus mutations in these regions can cause various diseases, *e.g.* spinal muscular atrophy, a common and leading cause of infant mortality, is shown to occur due to mutations in splicing region [23, 24]. Also, disruption in splicing of microtubule-associated protein tau (MAPT) gene can lead to neurodegenerative disorders [25] such as dementia, Alzheimer and Parkinsonism associated with chromosome 17 (FTDP-17) [26]. Furthermore, exon skipping in medium-chain acyl-CoA dehydrogenase gene can lead to severe enzyme deficiency causing metabolic disorders, like hypoglycemia [27, 28].

There are a number of functionally relevant non-coding RNAs discovered in last decades, primarily including piRNAs (PIWI-interacting RNAs), miRNA (microRNAs), siRNAs (small interfering RNAs), snoRNAs (small nucleolar RNAs), snRNAs (small nuclear RNAs), long noncoding RNAs (lncRNAs), *etc.* [29, 30]. These are involved in a multitude of diseases and regulation, for example, piRNA causes repression of transposable elements involved in genetic instability and is also associated with regulation of different cancers [31, 32]. Similarly, snoRNAs in human have been shown to be involved in neuro-developmental genetic disorder (due to the inefficient expression of C/D box snoRNAs) and cancer development [33]. Recently, it has been found that long noncoding RNAs also play a role in gene regulation by competing for endogenous RNAs (ceRNAs) and have severe pathological implications [34, 35]. miRNAs are another class of abundant small non-coding RNAs which have been implicated in various diseases. They are involved in glucose homeostasis [36], cancer development and progression [37], and also in neurodegenerative diseases [38, 39]. Modifications of RNA molecules are also crucial for the regulation of biological processes and have been involved in diseases [40]. Similar to DNA and histone modifications, RNA post-translational modifications represent a layer of epigenetic regulation. Methylation of adenosine at the N6 position (m6A) is an abundant mark in eukaryotic mRNA [41]. The modification of m6A is involved in a variety of biological processes and has been linked to several human diseases [42].

One of the tools which is very critical in unraveling the roles of RNA in health and diseases is the large-scale sequencing. Sequencing has revolutionized the field of biomedical research by analyzing clinically relevant samples at an unprecedented resolution than ever, helping in identifying new targets, regulators, biomarkers, and now it is even possible to interrogate entire genomes. For example, high-throughput sequencing has shown that the number of splicing sites in the human transcriptome is far more than identified earlier [43]. Similarly, sequencing the samples from breast cancer patients has identified several piRNAs which are differentially regulated in tumors compared to normal tissues [44]. Recent sequencing technologies coupled with antibody-mediated capture were also able to accurately map and quantify the m6A epigenetic modification (by m6A-seq) [45]. The analysis of the m6A distribution along the genome suggested that this mark could be involved in mediating splicing mechanisms, since transcripts with multiple isoforms were found to be enriched in m6A compared to single-isoform genes.

As it is commonly described, we are in the 2nd generation of sequencing and progressing rapidly towards the 3rd generation [46]. The pioneer sequencing methods developed by Frederick Sanger [47, 48] and Allan Maxam, Walter Gilbert [49] are often regarded as the first generation methods. While the chain termination strategy (Sanger sequencing method) has been used to sequence the first human genome [50] and is still considered as the 'gold standard' for sequencing, development of novel and efficient ways for creating clonal DNA population, less labor-intensive protocols, and technological advancement paved the way for second-generation sequencing methods [51 - 53]. These methods have not only reduced the cost and time of sequencing but, have unprecedentedly increased the coverage and resolution. All the second generation methods rely on clonal amplification step, either directly on glass flow-cells [54] or beads/emulsions [55, 56] or tubes (DNA nanoballs [57, 58],). Then the incorporation of nucleotides and detection is performed either by commonly used SBS (sequencing by synthesis) method, or by SBL (sequencing by ligation) methods [52, 53]. With advanced technological developments, now it is possible to get sequencing information even from the single DNA or RNA molecule [59, 60]. These single molecule sequencing methods are often regarded as the third generation sequencing methods and include strategies which are either based on zero-mode waveguide [61] (ZWV, SMRT sequencing [62, 63], PacBIO) or nanopores (Nanopore sequencing [64], Oxford Nanopore).

RNA sequencing (RNA-seq) [65] enables us to obtain sequence information from *in vitro* or *in vivo* RNA samples by implementing some additional molecular biology steps to prepare sequencing libraries compared to the standard DNA sequencing. Here RNAs are at first converted into complementary DNA (cDNA), to which universal sequencing adaptors are added and then the samples are often amplified either exponentially or linearly prior to the sequencing. Even prior to the development of high-throughput sequencing approaches, RNA molecules were quantified as well as sequenced using gene-specific fluorescent probes and sequencing *in situ* (FISSeq) [66, 67]. Though one of the pioneers, it suffered from very low throughput, labor-intensive and expensive protocols, and low sensitivity. Current RNA-seq methods have distinct advantages over earlier approaches to quantify transcriptomes and provide much better resolution for studying the gene expression profiles. As several laboratories are working on RNA-seq, the complete workflow is under continuous development for the technologies, efficient library preparation protocols, and data analysis, allowing to obtain even single-cell expression profiles at a remarkable resolution [68 - 70]. Furthermore, similar to single-molecule DNA sequencing, now it is also possible to sequence directly single RNA molecules without the requirement of the intermittent cDNA [71 - 73], overcoming the shortcomings of cDNA generations.

PRINCIPLE AND TECHNOLOGY

As previously mentioned, state-of-the-art of RNA-seq methods involve several steps like isolating RNAs from cells, enriching desired RNA subtypes, cDNA synthesis, appending sequencing adaptors, and various quality control steps prior to the sequencing [74 - 76]. Depending on the experimental needs of the researchers, within each of these steps, several sub-steps can be added. All these steps can be broadly classified into three parts (Fig. **1**); a) pre-sequencing treatment which includes steps from RNA isolation to the sequencing libraries preparation, b) sequencing, and c) post-sequencing bioinformatics analysis. Each of these steps includes several quality control procedures to ensure successful sequencing.

a. Pre-Sequencing:

The first crucial step which significantly contributes to the quality of sequencing is the isolation of RNA from biological samples. These samples contain a large variety of RNAs with the abundance in ribosomal RNA (rRNA), genomic DNA, and nucleases, therefore, it is quite important to maintain the integrity and quality of RNA during the isolation. The second crucial aspect is to enrich the desired RNA to be able to sequence them efficiently. Around ~80-90% of total RNA in the cell is estimated to be rRNA and thus it is important to remove them in order to achieve substantial sequencing depths and resolution [77, 78]. One of the easiest ways for the transcriptome studies interested in expressed mRNAs is using an oligo-dT containing solid supports (such as magnetic beads) to selectively enrich polyA-tailed mRNA [79]. Alternatively, rRNA depletion methods, like using exonucleases, can be employed to selectively degrade the non-capped rRNA while protecting the expressed mRNAs. Nowadays, there is a large variety of commercial RNA isolation kits available (from different suppliers like, ThermoScientific, Qiagen, Sigma, New England Biolabs, *etc.*) which are efficient, easy-to-use, and well suited to the isolation of different type of RNAs [80, 81]. While small RNAs like siRNAs, piRNAs, miRNAs can be sequenced completely by directly appending the sequencing adaptors, mRNAs are reasonably long and need to be fragmented to the appropriate size for the sequencing (~100-600 nucleotide long) [68, 74, 75]. This can be achieved by simple methods like limited RNA degradation, chemical treatment (*e.g.* using Zinc ions), and nebulization. RNAse III which cleaves dsRNA can also be used to fragment RNAs, however it produces less homogenous fragments than by Zn^{2+} ions [82]. Fragmentation can also be done at the cDNA level using sonication or DNAse I treatment. It has been shown that while cDNA fragmentation is more biased to generate 3'-end fragments of the gene (due to polyA enrichment) [83], RNA fragmentation covers the transcripts rather homogenously, but tends to deplete the sequence information at the 5' and 3' extremities [65, 84]. Illumina has also introduced a 'tagmentation' strategy where DNA is directly appended with adaptors during the fragmentation using a transposase enzyme pre-loaded with adaptor sequences [85]. This reduces the number of steps and biases involved in the library preparation process.

After fragmentation, the next step is to add adapter sequences that are complementary to the oligonucleotides bound on the flow-cell and needed to perform the sequencing cycle. These adaptors can be appended either by ligation or by step-wise PCR amplification, however, each of these steps has its own biases. RNA adaptor ligation is most commonly done by subsequently adding pre-adenylated 3' adaptor using Rnl2 [truncated RNA ligation 2] and then appending 5'adaptor using T4 RNA ligase 1 [Rnl1] [86]. This last step is followed by reverse transcription and PCR amplification to prepare the final library. While this process is the most commonly used, ligases have substrate sequence preferences, usually at the ligation site, which are prone to introduce biases in the sequence coverage and often less efficient with the low amount of RNAs [87, 88]. Similarly, the PCR steps are also amenable to biases [74, 75]. For example, due to the efficient amplification of GC-neutral regions, GC and AT-rich sequences are often underrepresented. Even DNA polymerases have differences in amplification efficiencies and uniform sequences coverage [89]. Amid these biases, linear amplification methods have been developed using T7 RNA polymerase or phi-29 DNA polymerase. The T7 RNA polymerase-based system [90], relies on attaching T7 promoter region to the fragmented sample, which then is isothermally transcribed in several copies using first strand (after RT) as a template. These RNAs are converted back to cDNA for the preparation of sequencing libraries. The phi-29 DNA polymerase performs isothermal rolling circle amplification where DNA fragments can be converted into a circular template by attaching 'dumb-bell' adaptors [91]. While both these linear amplification methods bring the advantage of avoiding PCR over-amplification of the sample. T7 polymerase method is prone to generate chimeric reads and moreover, phi-29 DNA polymerase has high error rates [74, 90].

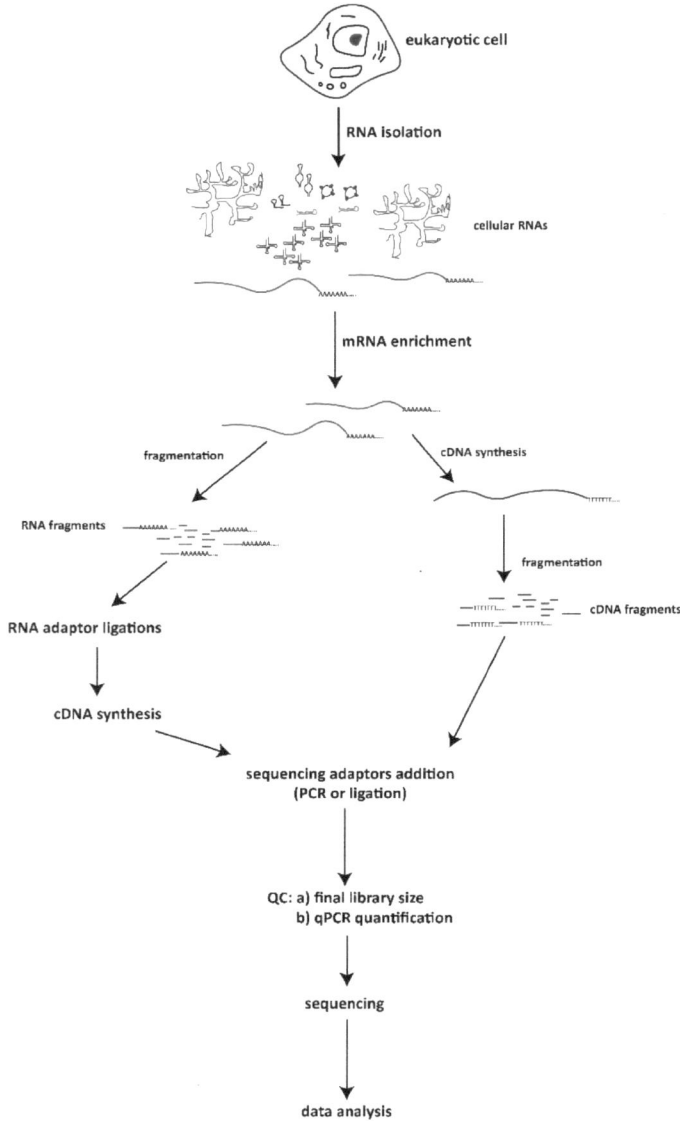

Fig. (1). Schematic representation of various steps involved in the RNA sequencing.

b. Sequencing:

Depending on the need of the researcher and the sample type, there are different next-generation sequencing platforms available. Out of these, Illumina® Sequencing [54, 92] leads most of the sequencing studies mainly due to its flexibility, low cost, easy-to-adapt protocols, and a wide range of available sequencing formats with reasonably high sequencing depths. Illumina uses reversible dye terminator chemistry to read each base (sequencing-by-synthesis approach [54]) of the sequence in a massively parallel way. Here, adaptor containing ssDNA molecules is hybridized on the flow-cell surface and amplified by bridging with the nearby complementary sequences (bridge amplification). This generates millions of cluster on the flow-cell, each containing the clonal copies of the original input DNA. Then using 3' reversibly blocked nucleotides labeled with fluorescent moieties, polymerase and the universal Illumina primer (Read1 or SBS3) each base of the sequence is read. After each incorporation, fluorescent image of flow cell is recorded and the 3' blocks are chemically removed to catalyze next incorporation. These steps are then iterated to read the complete amplicon. For the longer amplicons, DNA molecules can also read from the other end (3'-end) by 'pair-end reading' strategy. Here, after reading the sequence in the forward direction (using Read1, 5'-end), the DNA strands are flipped by binding to the complementary adaptor on the flow-cell and the strand is filled to generate the template for the sequencing from 3' direction. Then in a similar way as Read 1, sequencing cycle is performed using Read 2 primer. The sequence information of both the reads (Read 1 and Read 2) from the same cluster are merged together to extract data from the bigger amplicons. Illumina offers different sequencing machines [92] which differ in the throughput, length of the reads, imagining strategies, and sequencing times. Out of these, HiSeq and Miseq have been the workhorse for the sequencing where former produces very high quality billions of reads while the latter is more cost-effective with an output of up to 50 million reads. MiSeq sequencers have also low-output flow-cells, for example, nanoMiseq and microMiseq, where data from up to 4 million sequences can be obtained within a day with a price comparable to the sequencing of 96 clones using Sanger sequencing strategy. While the sequencing length is only up to ~300bp for such Miseq runs compared to ≥1000 bp by Sanger sequencing, such cost-effective options provide a simple and quick way to check the sequencing samples.

Unlike Illumina's modified bases for the sequencing, 'Ion Torrent™' sequencing technology [93, 94] has taken an entirely new approach by measuring the change in a physical property during the nucleotide incorporation. Ion Torrent uses a flow-cell with pre-defined wells, each connected to a pH sensor from the bottom. Whenever a base is incorporated, the pH of the well changes due to the release of the proton H[+] and thus measured by pH sensor. This makes the whole process faster and simpler as it involves only natural nucleotides. Here also prior to the sequencing, individual DNA sequences are amplified on a bead using emulsion PCR [94] and then added to the flow-cell such that each well contains only one bead. While this technology significantly reduced the time and cost, it is amenable to errors when homopolymers are sequenced [95].

Both of the above-mentioned technologies require clonal amplification ('clusters' or 'on beads') of DNA molecules prior to the sequencing and the fluorescence signal is generated from thousands of clonal sequences in a cluster during each incorporation. However, PacBio's SMRT (Single Molecule Real time sequencing) [63, 96] technology omits the pre-amplification requirement by the ability to sequence single DNA molecules in a massive parallel manner. This technology uses zero-mode waveguides (ZMW) [61], a cavity with less than a nanometer in diameter, to reduce the background signal by several orders of magnitude. Thus it is able to record fluorescence signal of each nucleotide incorporation from a single DNA molecule, copied by the polymerase fixed at the bottom of these cavities. Like other techniques, PacBio SMRT's flow-cell contains thousands to a million of ZMWs and several flow-cells can be sequenced in parallel. Library preparation protocol involves the conversion of input DNA into a circular template (SMRT Bells) by ligating dumb-bell adaptors for rolling circle amplification. This helps in compensating for the high error rates (~14%) by reading the same sequence several times in ZMW to generate a consensus sequence with high accuracy (~99.99%), comparable (or even more) than existing other sequencing platforms. One of the big advantages here is the possibility to measure polymerase kinetics which can be used to identify modified bases, *e.g.* 5-mC, 5-hmC, as each of them will have a different incorporation rate. This allows to obtain the epigenetic information of the sample simultaneously during the sequencing [97].

Another approach that is quite different from the other sequencing platforms is 'sequencing by ligation' [98] from ABI SOLiD platform. Here the sequence information from the DNA fragments is deciphered by sequential ligation of short di-base DNA strand (8 nucleotides long) in a combinatorial way. The input DNA is pre-amplified on beads using emulsion PCR and bound to the glass flow-cell. Then four fluorescently labeled di-base ligation probes, containing all the combination of dinucleotides on one end (16 combinations with four different colors; each color corresponds to four different combinations) are added along with ligase enzyme. After each ligation step the signal is measured, the fluorescent part of the probe is cleaved off, and the consecutive ligation steps are carried out to reveal the sequence information of every 5th base using the fluorescence signal associated with each combination. The whole ligation process is then repeated, each time using probes that are offset by one base (n-1, n-2, n-3, n-4), thus revealing the information from the complete sequence. Though ligation process here utilizes two bases at the junction for accurate identification of the correct base compared to other methods where they recognize only one base at a time during polymerization, the technology is hampered by palindromic repeats in the sequences [99]. Nevertheless, the sequencing by ligation approach has been used in several clinical and non-clinical studies [100 - 102].

A different approach based on measuring pyrophosphate released during the polymerization process, 'Pyrosequencing or 454 Sequencing' [103], was also developed during the initial periods of sequencing platform development. However, due to limited use and strong competition from other platforms, pyrosequencing did not have commercial success.

One of the promising single molecule sequencing methods available these days is the strategy based on nanopores from Oxford Nanopore technologies [64, 104]. The technology exploits pores of nm size [9, 10] either natural *e.g.* α-hemolysin or artificial which are submerged in an electrolyte solution allowing the measurement of the electric field [105]. The single DNA or RNA fragments are bound to the processive enzymes and captured on the nanopores grafted on solid support. Then the fragments are passed through the pores, and the change in electric field is measured with each nucleotide passed. As the composition of each nucleotide gives a characteristic change in the electric field, it can be used for the base calling. The complete process is pretty rapid, low-cost, and handy as Oxford Nanopore technologies have also developed a low-throughput device (MinION [104],) which is as small as USB sticks and can be carried anywhere for the sequencing. Even though the nanopore methods still have higher error rates compared to other platforms, measuring the electric field based on the physical properties of the nucleotide opens up the possibility to sequence the samples containing modified bases [71, 106].

c. Post-Sequencing:

All these massive parallel sequencing methods generate an enormous volume of data which needs to be handled efficiently to obtain useful information [107, 108]. The foremost step is to do base calling [109] which converts raw signal data [*e.g.,* fluorescence intensities] into the correct corresponding base. While most of the sequencing platforms perform base-calling for the users, the algorithms can also be installed for the in-house analysis pipelines. The reliability of the called base can be checked for 'phred quality score' [110, 111] which measures the probability of errors in base calling (*e.g.,* $Q = -10 \log P$, where P is the error probability). The error in measurement comes from imperfect signal intensities, too many clusters nearby, and background noise. After base calling, the sequences are presented as simple readable sequence file, *e.g. fastq* files which contain called bases, cluster ids, index information, and the quality scores. Often after this step, low quality sequences are filtered out, adaptors sequences are trimmed and the remaining sequences are aligned to the reference genomes. There are a couple of helpful software packages like Burrows-Wheeler aligner (BWA) [112], Bowtie [113], string graph assembler (SGA) [114, 115], *etc.* which can be included in custom analysis pipelines to align the sequencing reads. The alignment step is also prone to errors because of the difficulty in unequivocally align a read to a perfect genome location, therefore, a mapping quality score should also be calculated. After these basic pre-processing steps, the high quality sequencing reads can be analyzed in different ways based on the requirements of the researchers. As the majority of the sequencing platforms relies on amplification of the starting material, one of the main concern of the differential gene expression analysis is the biases in amplification efficiencies. Such biases can change the representation of the actual number of the expressed genes and, are even more problematic for lowly expressed genes [116]. The amplification step is indeed required as often RNA from the biological samples are low in concentrations. However, the over amplification mask the actual differences in sequencing read numbers due to PCR duplicates. To overcome such problems a stretch of random nucleotides (often 6-8 nt) can be introduced during reverse transcription step (therefore prior to PCR amplification), such that each cDNA produced will contain a unique sequence combination called as 'unique molecular identifiers (UMI)' [117]. Analyzing the number of UMI per gene provides more accurate quantification of gene expression. Even though the UMIs are useful in normalizing the amplification biases, they are also prone to PCR and sequencing errors which generate false UMIs with very few reads. These problems are well characterized and, therefore exhaustive algorithms are developed to overcome it by collapsing similar UMIs [118, 119].

QUALITY CONTROL STRATEGIES IN THE NEXT-GENERATION SEQUENCING

In order to achieve high-quality sequencing reads, quality control (QC) steps should be implemented at different stages of the sequencing protocol (Fig. 1). Such QC steps can be applied directly from the RNA isolation, *e.g.*, controlling its integrity. Another critical factor that affects the quality of the sequencing is the reliable quantification of the actual sequencable molecules in the library. While fluorescence based measurements (*e.g.* Qubit™) provide accurate quantification of the library concentrations, it does not guarantee that all the measured molecules carry the specific adaptors required for the sequencing. Therefore two important measures should be taken into account; first checking the size of the sample by sensitive electrophoretic methods (Bioanalyzer, TapeStation, LabChip GX), and second is the quantification by qPCR using sequencing adaptor-specific primers. Using standard qPCR-based quantifications kits (like Kappa Kit from Roche, NEBNext kit from New England Biolabs), the exact concentration of sequenceable molecules in the libraries can be quantified which helps in the optimum loading of the DNA fragments on the sequencer. It is very critical to have the ideal loading on the sequencer as both under-loading and over-loading create noisy signals and dramatically affect the sequencing outputs.

For the RNA expression analysis, it is also important to account for the sequence data variations due to different sequencing platforms as well as batch-effects, by implementing reliable positive controls as quantitative standards. These variations come from different library preparation protocols, RNA quality of each batch, sequencing strategy, and other technical noises. To standardize the process, External RNA controls Consortium, together with National Institute of Standards and Technology (ERCC) [120] has developed a set of 92 control RNAs of varying length and GC content which can be spike-in with each sample and sequenced together [121]. These RNAs are polyadenylated with known sequences and having different concentrations, therefore act as quantitative standards for the better quantification of expressed cellular genes and sequencing coverage.

ADVANCED TECHNIQUES

Single Cell RNA Sequencing

While conventional RNA transcriptomic studies from bulk samples (more than 1 cell) provide a good overview of the average expression profile in the sample, they often mask specific cell information, important in understanding cell-to-cell variability. There is often heterogeneity in gene expression of phenotypically similar cell populations. Also within a tissue, there are different cell phenotypes with specific functions and surface makers, *e.g.* immune cell repertoire [122]. In order to capture such variabilities, in the last few years, several research groups have developed single-cell RNA sequencing (scRNA-seq) protocols [123 - 125]. There are several hurdles which needed to overcome for the scRNA-seq like the reliable isolation of single cells from complex samples, low amount of RNA from single cells (~1 pg), the poor efficiency of cDNA synthesis [126], and scaling up to capture more cells [127]. The scRNA-seq protocols differ from each other in the number of cells analyzed, the technological strategy used (plates, droplets), sequencing depth, number of genes per cell, sensitivity in gene detection, and accuracy in quantification [124, 128]. Other variations include amplification strategy used, cDNA coverage, use of UMIs, *etc.* While a lot of these scRNA-seq protocols use micro-well plates, *e.g.* Smart-seq2 [129], CEL-seq [130], MARS-seq [131], BAT-seq [132], which limit the number of cells analyzed, protocols based on microfluidic chips, *e.g.* C1 system (Fluidigm) CEL-seq2 (C1) [133], can handle a couple of thousands of cells (up to 8000) [134]. With the inclusion of nano-liter droplets along with drop-specific barcodes on beads, *e.g.* in Drop [135] (Fig. **2A**), Drop-Seq [136], Chromium system [137], the cell-throughput has further increased to 10^5 cells. These drop-based methods are still under continuous development and they will increase the capacity even further. More recently, a plate-based method, SPLiT-seq [138], which exploits fixed cells and randomly added barcodes has increased the cell-limit by another order of magnitude (10^6 cells). All these protocols have been systematically benchmarked using ERCC control RNAs for their accuracy (precise estimation of the RNA concentration) and sensitivity *i.e.* the lowest detection limit [139, 140]. While scRNA-seq protocols were found to be more accurate and sensitive than the bulk sequencing methods, within them there are huge variations. Nevertheless, with extensive efforts in improving these protocols, researcher are now able to examine thousands of cells at a single cell resolution to address cellular heterogeneity with much more accuracy and efficiency than ever before.

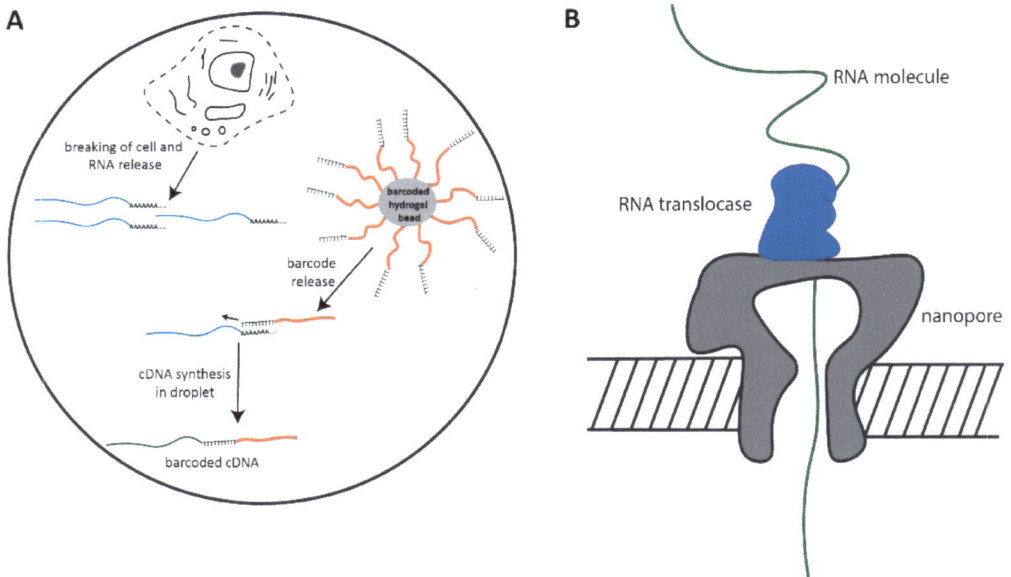

Fig. (2). Examples of advance technological development in the RNA sequencing. **A.** Single-cell sequencing strategy based on encapsulation of barcoded hydrogel beads in nano-litre droplets [135]. Here hydrogel beads containing unique sequence barcode (with >10^8 copies) are co-encapsulated with single cells in the droplets. Cell lysis, barcode release and the reverse transcription are performed inside the droplets to produce barcoded cDNAs, thus enabling to extract single-cell sequence information. **B.** Single RNA molecule sequencing strategy based on nanopores [71]. Here RNA bound to an RNA translocase enzyme is trapped over nanopore and fixed on the solid-surface. RNA is then passed through the pore and the change in electric potential is measured for each nucleotide.

Direct RNA Sequencing

Though most of the RNA-seq and scRNA-seq methods have revealed a vast and previously unknown wealth of transcriptome information, assisting in the better understanding of cell-to-cell variability, they are still inefficient to characterize splicing junctions and epigenetic modifications [65, 70, 124]. One of the steps which limits these sequencing approaches is the conversion of RNA to cDNA and the later amplification steps. cDNA synthesis is known to be inefficient, error prone due to template switching and furthermore downstream steps like amplification or ligation steps introduce biases. To overcome these issues, Ozolak *et al.* [73] demonstrated direct sequencing of RNA on a Helicos sequencer [141] by exploiting the reverse transcriptase activity of DNA polymerases. Later on, Mamanova *et al.* developed FRT-seq method [142] where they used a reverse transcriptase to sequence RNA molecules directly on Illumina flow-cell surface. Similar to the above mentioned method, here too the sequence information is decoded using the sequencing-by-synthesis approach. While both of these approaches have overcome the amplification biases while keeping original RNA strand information, they still rely on copying the RNA molecules limiting their ability to decipher epigenetic information. Also, due to shorter read lengths both these methods are not suited for revealing splicing junctions. To overcome these limitations, Oxford Nanopore technology has developed direct single-molecule RNA sequencing using nanopore omitting the requirement of RNA to cDNA conversion [71] (Fig. **2B**). While the method is still under development and yet to be able to compete with other existing strategies, it possesses great potential in reading long genes with low biases and at the same time revealing the base modification information required for the epigenetic analysis.

APPLICATIONS

RNA-seq has revolutionized the biomedical field by bringing unprecedented insights into health, diseases, and diagnostics. RNA defects have been implicated in many diseases, for example, malfunction in RNA splicing events has serious consequences and can lead to the development of various diseases [143]. RNA-seq has enabled detection of such splice events [144, 145], helped in the *de novo* prediction of splice variants [146], and also made it possible to identify aberrant splicing regions in cancer cells [147]. RNA-seq has also helped in identifying rare non-exonic variants related to neurologic dysfunction (caused by TIMMDC1 chaperon protein involved in the respiratory complex) [148]. Similarly, RNA-seq has accelerated the detection and identification of single nucleotide polymorphisms (SNPs), which are associated with a wide-range of diseases [149, 150]. For example, Kang *et al.* [149], using RNA-seq data from several unrelated European individuals, were able to identify enriched SNPs associated with immune-related diseases.

With the advent of the single-cell RNA sequencing (scRNA-seq) approaches several critical issues have been resolved, like differentiating and identifying novel cell types, quantifying gene-expression heterogeneity and understanding the developmental processes at the single-cell level. For example, scRNA-seq has enabled to confirm the expression of several mutants involved in acute myeloid leukemia [151]. Tissues of the central nervous system (CNS) are often quite heterogeneous with complex cell-types and reasonable low abundance of mRNAs, therefore studying them at single-cell level provides better insight for identifying biomarkers and understanding disease state [152]. For example, studying mouse brain cells by scRNA-seq deciphered dormant neural stem cell in case of 'brain ischemia' [153]. This study also revealed the influence of signaling molecule interferon (IFN)-γ for the activation from dormant to activated cells. scRNA-seq has also greatly contributed to organ development research. In a study using three different scRNA-seq protocols, the glial neurotrophic growth factor (GDNF), an important factor for early kidney development, was found to be produced mainly by stromal cell against the previously thought mesenchyme nephron progenitors [154]. scRNA-seq approaches have been also extensively used for studying immune cells development and heterogeneity [155]. Recent trend in the single-cell studies is to couple different sequencing strategies to simultaneously obtain information about genome, transcriptome epigenome and proteome, all at the single cell level. For instance scRNA-seq coupled with specific cell-surface antibodies detection [156] can decipher both transcriptome and phenotype at the same time, providing a more powerful way for the identification of cell-subtypes within complex samples like human tumor specimens.

CONCLUDING REMARKS

In the last few years, with rapid technological developments, deciphering the transcriptome complexity has been far more cheap and efficient. The RNA sequencing has produced a wealth of data revealing novel roles of RNA in diseases, aiding the biomarker development and contributing to the fundamental understanding of the biological process at the molecular level. There have been massive improvements in the RNA-seq protocols overcoming biases and making the process more robust and reproducible. With such improvements, even the transcriptome at a single-cell level can be sequenced. Even though the number of cells and sequencing depth in each cell is not ideal yet, the scRNA-seq methods have already enabled addressing the cell-to-cell heterogeneity and differential gene expressions pretty well. All this new information strongly demands the development of better strategies to deal with huge data-sets and make them comparable to each other. Apart from dealing with the amount of data, strong statistical approaches are also required to decipher the correct information from noisy sequencing data, especially from scRNA-seq studies.

CONSENT FOR PUBLICATION

Not applicable.

CONFLICT OF INTEREST

The author confirms that this chapter contents have no conflict of interest.

ACKNOWLEDGEMENTS

Authors thank Dr. Philippe Nghe and Prof. Dr. Andrew D. Griffiths, ESPCI Paris for their support.

REFERENCES

[1]   Erdmann VA, Barciszewska MZ, Hochberg A, de Groot N, Barciszewski J. Regulatory RNAs. Cell Mol Life Sci 2001; 58(7): 960-77.
[http://dx.doi.org/10.1007/PL00000913] [PMID: 11497242]

[2]   He L, Hannon GJ. MicroRNAs: small RNAs with a big role in gene regulation. Nat Rev Genet 2004; 5(7): 522-31.
[http://dx.doi.org/10.1038/nrg1379] [PMID: 15211354]

[3]   Lioliou E, Romilly C, Romby P, Fechter P. RNA-mediated regulation in bacteria: from natural to artificial systems. N Biotechnol 2010; 27(3): 222-35.
[http://dx.doi.org/10.1016/j.nbt.2010.03.002] [PMID: 20211281]

[4]   Serganov A, Patel DJ. Ribozymes, riboswitches and beyond: regulation of gene expression without proteins. Nat Rev Genet 2007; 8(10): 776-90.
[http://dx.doi.org/10.1038/nrg2172] [PMID: 17846637]

[5]   Waters LS, Storz G. Regulatory RNAs in bacteria. Cell 2009; 136(4): 615-28.
[http://dx.doi.org/10.1016/j.cell.2009.01.043] [PMID: 19239884]

[6]   Guil S, Esteller M. RNA-RNA interactions in gene regulation: the coding and noncoding players. Trends Biochem Sci 2015; 40(5): 248-56.
[http://dx.doi.org/10.1016/j.tibs.2015.03.001] [PMID: 25818326]

[7]   Lau NC, Lai EC. Diverse roles for RNA in gene regulation. Genome Biol 2005; 6(4): 315.
[http://dx.doi.org/10.1186/gb-2005-6-4-315] [PMID: 15833133]

[8]   Breaker RR. Riboswitches and the RNA world. Cold Spring Harb Perspect Biol 2012; 4(2): 4.
[http://dx.doi.org/10.1101/cshperspect.a003566] [PMID: 21106649]

[9]   Bagheri S, Kashani-Sabet M. Ribozymes in the age of molecular therapeutics. Curr Mol Med 2004; 4(5): 489-506.
[http://dx.doi.org/10.2174/1566524043360410] [PMID: 15267221]

[10]   Nissen P, Hansen J, Ban N, Moore PB, Steitz TA. The structural basis of ribosome activity in peptide bond synthesis. Science 2000; 289(5481): 920-30.
[http://dx.doi.org/10.1126/science.289.5481.920] [PMID: 10937990]

[11]   Scott WG. Ribozymes. Curr Opin Struct Biol 2007; 17(3): 280-6.
[http://dx.doi.org/10.1016/j.sbi.2007.05.003] [PMID: 17572081]

[12]   Valadkhan S. The spliceosome: a ribozyme at heart? Biol Chem 2007; 388(7): 693-7.
[http://dx.doi.org/10.1515/BC.2007.080] [PMID: 17570821]

[13]   Zhang SJ, Chen X, Li CP, *et al.* Identification and Characterization of Circular RNAs as a New Class of Putative Biomarkers in Diabetes Retinopathy. Invest Ophthalmol Vis Sci 2017; 58(14): 6500-9.
[http://dx.doi.org/10.1167/iovs.17-22698] [PMID: 29288268]

[14]   Zhang ZC, Guo XL, Li X. The novel roles of circular RNAs in metabolic organs. Genes Dis 2017; 5(1): 16-23.
[http://dx.doi.org/10.1016/j.gendis.2017.12.002] [PMID: 30258930]

[15]   Amaral PP, Mattick JS. Noncoding RNA in development. Mamm Genome 2008; 19(7-8): 454-92.
[http://dx.doi.org/10.1007/s00335-008-9136-7] [PMID: 18839252]

[16]   Barta A, Jantsch MF. RNA in Disease and development. RNA Biol 2017; 14(5): 457-9.
[http://dx.doi.org/10.1080/15476286.2017.1316929] [PMID: 28402218]

[17]   Mattick JS. The central role of RNA in human development and cognition. FEBS Lett 2011; 585(11): 1600-16.
[http://dx.doi.org/10.1016/j.febslet.2011.05.001] [PMID: 21557942]

[18]   Cooper TA, Wan L, Dreyfuss G. RNA and disease. Cell 2009; 136(4): 777-93.
[http://dx.doi.org/10.1016/j.cell.2009.02.011] [PMID: 19239895]

[19]   Hüttenhofer A, Schattner P, Polacek N. Non-coding RNAs: hope or hype? Trends Genet 2005; 21(5): 289-97.
[http://dx.doi.org/10.1016/j.tig.2005.03.007] [PMID: 15851066]

[20]   Palazzo AF, Lee ES. Non-coding RNA: what is functional and what is junk? Front Genet 2015; 6: 2.
[http://dx.doi.org/10.3389/fgene.2015.00002] [PMID: 25674102]

[21]   Black DL. Mechanisms of alternative pre-messenger RNA splicing. Annu Rev Biochem 2003; 72: 291-336.
[http://dx.doi.org/10.1146/annurev.biochem.72.121801.161720] [PMID: 12626338]

[22]   Faustino NA, Cooper TA. Pre-mRNA splicing and human disease. Genes Dev 2003; 17(4): 419-37.
[http://dx.doi.org/10.1101/gad.1048803] [PMID: 12600935]

[23]   Lorson CL, Hahnen E, Androphy EJ, Wirth B. A single nucleotide in the SMN gene regulates splicing and is responsible for spinal muscular atrophy. Proc Natl Acad Sci USA 1999; 96(11): 6307-11.
[http://dx.doi.org/10.1073/pnas.96.11.6307] [PMID: 10339583]

[24]   Wee CD, Kong L, Sumner CJ. The genetics of spinal muscular atrophies. Curr Opin Neurol 2010; 23(5): 450-8.
[http://dx.doi.org/10.1097/WCO.0b013e32833e1765] [PMID: 20733483]

[25]   Park SA, Ahn SI, Gallo JM. Tau mis-splicing in the pathogenesis of neurodegenerative disorders. BMB Rep 2016; 49(8): 405-13.
[http://dx.doi.org/10.5483/BMBRep.2016.49.8.084] [PMID: 27222125]

[26]   Wszolek ZK, Tsuboi Y, Ghetti B, Pickering-Brown S, Baba Y, Cheshire WP. Frontotemporal dementia and parkinsonism linked to chromosome 17 (FTDP-17). Orphanet J Rare Dis 2006; 1: 30.
[http://dx.doi.org/10.1186/1750-1172-1-30] [PMID: 16899117]

[27]   Nielsen KB, Sørensen S, Cartegni L, *et al.* Seemingly neutral polymorphic variants may confer immunity to splicing-inactivating mutations: a synonymous SNP in exon 5 of MCAD protects from deleterious mutations in a flanking exonic splicing enhancer. Am J Hum Genet 2007; 80(3): 416-32.
[http://dx.doi.org/10.1086/511992] [PMID: 17273963]

[28]   Strauss AW, Powell CK, Hale DE, *et al.* Molecular basis of human mitochondrial very-long-chain acyl-CoA dehydrogenase deficiency causing cardiomyopathy and sudden death in childhood. Proc Natl Acad Sci USA 1995; 92(23): 10496-500.
[http://dx.doi.org/10.1073/pnas.92.23.10496] [PMID: 7479827]

[29]   Mattick JS, Makunin IV. Non-coding RNA. Hum Mol Genet 2006; 15(Spec No 1): R17-29.
[http://dx.doi.org/10.1093/hmg/ddl046] [PMID: 16651366]

[30]   Taft RJ, Pang KC, Mercer TR, Dinger M, Mattick JS. Non-coding RNAs: regulators of disease. J Pathol 2010; 220(2): 126-39.
[http://dx.doi.org/10.1002/path.2638] [PMID: 19882673]

[31]   Iwasaki YW, Siomi MC, Siomi H. PIWI-Interacting RNA. Annu Rev Biochem 2015; 84: 405-33.
[http://dx.doi.org/10.1146/annurev-biochem-060614-034258] [PMID: 25747396]

[32]   Moyano M, Stefani G. piRNA involvement in genome stability and human cancer. J Hematol Oncol 2015; 8: 38.
[http://dx.doi.org/10.1186/s13045-015-0133-5] [PMID: 25895683]

[33]   Williams GT, Farzaneh F. Are snoRNAs and snoRNA host genes new players in cancer? Nat Rev Cancer 2012; 12(2): 84-8.
[http://dx.doi.org/10.1038/nrc3195] [PMID: 22257949]

[34]   de Giorgio A, Krell J, Harding V, Stebbing J, Castellano L. Emerging roles of competing endogenous RNAs in cancer: insights from the regulation of PTEN. Mol Cell Biol 2013; 33(20): 3976-82.
[http://dx.doi.org/10.1128/MCB.00683-13] [PMID: 23918803]

[35]   Kartha RV, Subramanian S. Competing endogenous RNAs (ceRNAs): new entrants to the intricacies of gene regulation. Front Genet 2014; 5: 8.
[http://dx.doi.org/10.3389/fgene.2014.00008] [PMID: 24523727]

[36]   Deiuliis JA. MicroRNAs as regulators of metabolic disease: pathophysiologic significance and emerging role as biomarkers and therapeutics. Int J Obes 2016; 40(1): 88-101.
[http://dx.doi.org/10.1038/ijo.2015.170] [PMID: 26311337]

[37]   Hayes J, Peruzzi PP, Lawler S. MicroRNAs in cancer: biomarkers, functions and therapy. Trends Mol Med 2014; 20(8): 460-9.
[http://dx.doi.org/10.1016/j.molmed.2014.06.005] [PMID: 25027972]

[38]   Delay C, Mandemakers W, Hébert SS. MicroRNAs in Alzheimer's disease. Neurobiol Dis 2012; 46(2): 285-90.
[http://dx.doi.org/10.1016/j.nbd.2012.01.003] [PMID: 22285895]

[39]   Nunez-Iglesias J, Liu CC, Morgan TE, Finch CE, Zhou XJ. Joint genome-wide profiling of miRNA and mRNA expression in Alzheimer's disease cortex reveals altered miRNA regulation. PLoS One 2010; 5(2) e8898
[http://dx.doi.org/10.1371/journal.pone.0008898] [PMID: 20126538]

[40]   Jonkhout N, Tran J, Smith MA, Schonrock N, Mattick JS, Novoa EM. The RNA modification landscape in human disease. RNA 2017; 23(12): 1754-69.
[http://dx.doi.org/10.1261/rna.063503.117] [PMID: 28855326]

[41]   Horowitz S, Horowitz A, Nilsen TW, Munns TW, Rottman FM. Mapping of N6-methyladenosine residues in bovine prolactin mRNA. Proc Natl Acad Sci USA 1984; 81(18): 5667-71.
[http://dx.doi.org/10.1073/pnas.81.18.5667] [PMID: 6592581]

[42]   Zhao BS, Roundtree IA, He C. Post-transcriptional gene regulation by mRNA modifications. Nat Rev Mol Cell Biol 2017; 18(1): 31-42.
[http://dx.doi.org/10.1038/nrm.2016.132] [PMID: 27808276]

[43]   Pan Q, Shai O, Lee LJ, Frey BJ, Blencowe BJ. Deep surveying of alternative splicing complexity in the human transcriptome by high-throughput sequencing. Nat Genet 2008; 40(12): 1413-5.
[http://dx.doi.org/10.1038/ng.259] [PMID: 18978789]

[44]   Huang G, Hu H, Xue X, *et al.* Altered expression of piRNAs and their relation with clinicopathologic features of breast cancer. Clin Transl Oncol 2013; 15(7): 563-8.
[http://dx.doi.org/10.1007/s12094-012-0966-0] [PMID: 23229900]

[45]   Dominissini D, Moshitch-Moshkovitz S, Schwartz S, *et al.* Topology of the human and mouse m6A RNA methylomes revealed by m6A-seq. Nature 2012; 485(7397): 201-6.
[http://dx.doi.org/10.1038/nature11112] [PMID: 22575960]

[46]   Heather JM, Chain B. The sequence of sequencers: The history of sequencing DNA. Genomics 2016; 107(1): 1-8.
[http://dx.doi.org/10.1016/j.ygeno.2015.11.003] [PMID: 26554401]

[47]   Sanger F Coulson AR. A rapid method for determining sequences in DNA by primed synthesis with DNA polymerase. J Mol Biol 1975; 94(3): 441-8.
[http://dx doi.org/10.1016/0022-2836(75)90213-2] [PMID: 1100841]

[48]   Sanger F, Nicklen S, Coulson AR. DNA sequencing with chain-terminating inhibitors. Proc Natl Acad Sci USA 1977; 74(12): 5463-7.
[http://dx.doi.org/10.1073/pnas.74.12.5463] [PMID: 271968]

[49]   Maxam AM, Gilbert W. A new method for sequencing DNA. Proc Natl Acad Sci USA 1977; 74(2): 560-4.
[http://dx.doi.org/10.1073/pnas.74.2.560] [PMID: 265521]

[50]   Venter JC, Adams MD, Myers EW, *et al.* The sequence of the human genome. Science 2001; 291(5507): 1304-51.
[http://dx.doi.org/10.1126/science.1058040] [PMID: 11181995]

[51]   Church GM. Genomes for all. Sci Am 2006; 294(1): 46-54.
[http://dx.doi.org/10.1038/scientificamerican0106-46] [PMID: 16468433]

[52]   Metzker ML. Emerging technologies in DNA sequencing. Genome Res 2005; 15(12): 1767-76.
[http://dx.doi.org/10.1101/gr.3770505] [PMID: 16339375]

[53]   Metzker ML. Sequencing technologies - the next generation. Nat Rev Genet 2010; 11(1): 31-46.
[http://dx.doi.org/10.1038/nrg2626] [PMID: 19997069]

[54]   Bentley DR, Balasubramanian S, Swerdlow HP, *et al.* Accurate whole human genome sequencing using reversible terminator chemistry. Nature 2008; 456(7218): 53-9.
[http://dx.coi.org/10.1038/nature07517] [PMID: 18987734]

[55]   Margulies M, Egholm M, Altman WE, *et al.* Genome sequencing in microfabricated high-density picolitre reactors. Nature 2005; 437(7057): 376-80.
[http://dx.doi.org/10.1038/nature03959] [PMID: 16056220]

[56]   Wheeler DA, Srinivasan M, Egholm M, *et al.* The complete genome of an individual by massively parallel DNA sequencing. Nature 2008; 452(7189): 872-6.
[http://dx.doi.org/10.1038/nature06884] [PMID: 18421352]

[57]   Drmanac R, Sparks AB, Callow MJ, *et al.* Human genome sequencing using unchained base reads on self-assembling DNA nanoarrays. Science 2010; 327(5961): 78-81.
[http://dx.doi.org/10.1126/science.1181498] [PMID: 19892942]

[58]   Porreca GJ. Genome sequencing on nanoballs. Nat Biotechnol 2010; 28(1): 43-4.
[http://dx.doi.org/10.1038/nbt0110-43] [PMID: 20062041]

[59]   Gupta PK. Single-molecule DNA sequencing technologies for future genomics research. Trends Biotechnol 2008; 26(11): 602-11.
[http://dx.doi.org/10.1016/j.tibtech.2008.07.003] [PMID: 18722683]

[60]   Xu M, Fujita D, Hanagata N. Perspectives and challenges of emerging single-molecule DNA sequencing technologies. Small 2009; 5(23): 2638-49.
[http://dx.doi.org/10.1002/smll.200900976] [PMID: 19904762]

[61]   Zhu P, Craighead HG. Zero-mode waveguides for single-molecule analysis. Annu Rev Biophys 2012; 41: 269-93.
[http://dx.dci.org/10.1146/annurev-biophys-050511-102338] [PMID: 22577821]

[62]   Ardui S, Ameur A, Vermeesch JR, Hestand MS. Single molecule real-time (SMRT) sequencing comes of age: applications and utilities for medical diagnostics. Nucleic Acids Res 2018; 46(5): 2159-68.
[http://dx.doi.org/10.1093/nar/gky066] [PMID: 29401301]

[63]   Eid J, Fehr A, Gray J, *et al.* Real-time DNA sequencing from single polymerase molecules. Science 2009; 323(5910): 133-8.
[http://dx.doi.org/10.1126/science.1162986] [PMID: 19023044]

[64]   Clarke J, Wu HC, Jayasinghe L, Patel A, Reid S, Bayley H. Continuous base identification for single-molecule nanopore DNA sequencing. Nat Nanotechnol 2009; 4(4): 265-70.
[http://dx.doi.org/10.1038/nnano.2009.12] [PMID: 19350039]

[65]   Wang Z, Gerstein M, Snyder M. RNA-Seq: a revolutionary tool for transcriptomics. Nat Rev Genet 2009; 10(1): 57-63.
[http://dx.doi.org/10.1038/nrg2484] [PMID: 19015660]

[66]   Lee JH, Daugharthy ER, Scheiman J, *et al.* Fluorescent in situ sequencing (FISSEQ) of RNA for gene expression profiling in intact cells and tissues. Nat Protoc 2015; 10(3): 442-58.
[http://dx.doi.org/10.1038/nprot.2014.191] [PMID: 25675209]

[67]   Lee JH, Daugharthy ER, Scheiman J, *et al.* Highly multiplexed subcellular RNA sequencing in situ. Science 2014; 343(6177): 1360-3.
[http://dx.doi.org/10.1126/science.1250212] [PMID: 24578530]

[68]   Hrdlickova R, Toloue M, Tian B. RNA-Seq methods for transcriptome analysis. Wiley Interdiscip Rev RNA 2017; 8(1): 8.
[http://dx.doi.org/10.1002/wrna.1364] [PMID: 27198714]

[69]   Kukurba KR, Montgomery SB. RNA Sequencing and Analysis. Cold Spring Harb Protoc 2015; 2015(11): 951-69.
[http://dx.doi.org/10.1101/pdb.top084970] [PMID: 25870306]

[70]   Ozsolak F, Milos PM. RNA sequencing: advances, challenges and opportunities. Nat Rev Genet 2011; 12(2): 87-98.
[http://dx.doi.org/10.1038/nrg2934] [PMID: 21191423]

[71]   Garalde DR, Snell EA, Jachimowicz D, *et al.* Highly parallel direct RNA sequencing on an array of nanopores. Nat Methods 2018; 15(3): 201-6.
[http://dx.doi.org/10.1038/nmeth.4577] [PMID: 29334379]

[72]   Ozsolak F, Milos PM. Single-molecule direct RNA sequencing without cDNA synthesis. Wiley Interdiscip Rev RNA 2011; 2(4): 565-70.
[http://dx.doi.org/10.1002/wrna.84] [PMID: 21957044]

[73]   Ozsolak F, Platt AR, Jones DR, *et al.* Direct RNA sequencing. Nature 2009; 461(7265): 814-8.
[http://dx.doi.org/10.1038/nature08390] [PMID: 19776739]

[74]   Head SR, Komori HK, LaMere SA, *et al.* Library construction for next-generation sequencing: overviews and challenges. Biotechniques 2014; 56(2): 61-64, 66, 68 passim.
[http://dx.doi.org/10.2144/000114133] [PMID: 24502796]

[75]   van Dijk EL, Jaszczyszyn Y, Thermes C. Library preparation methods for next-generation sequencing: tone down the bias. Exp Cell Res 2014; 322(1): 12-20.
[http://dx.doi.org/10.1016/j.yexcr.2014.01.008] [PMID: 24440557]

[76]   Podnar J, Deiderick H, Huerta G, Hunicke-Smith S. Next-Generation Sequencing RNA-Seq Library Construction. Curr Protoc Mol Biol 2014; 106: 1-19.

[77]   Karpinets TV, Greenwood DJ, Sams CE, Ammons JT. RNA:protein ratio of the unicellular organism as a characteristic of phosphorous and nitrogen stoichiometry and of the cellular requirement of ribosomes for protein synthesis. BMC Biol 2006; 4: 30.
[http://dx.doi.org/10.1186/1741-7007-4-30] [PMID: 16953894]

[78]   O'Neil D, Glowatz H, Schlumpberger M. Ribosomal RNA depletion for efficient use of RNA-seq capacity. Curr Protoc Mol Biol 2013; 103: 1-18.

[79]   Aviv H, Leder P. Purification of biologically active globin messenger RNA by chromatography on oligothymidylic acid-cellulose. Proc Natl Acad Sci USA 1972; 69(6): 1408-12.
[http://dx.doi.org/10.1073/pnas.69.6.1408] [PMID: 4504350]

[80]   Morlan JD, Qu K, Sinicropi DV. Selective depletion of rRNA enables whole transcriptome profiling of archival fixed tissue. PLoS One 2012; 7(8)e42882
[http://dx.doi.org/10.1371/journal.pone.0042882] [PMID: 22900061]

[81]   Zhao W, He X, Hoadley KA, Parker JS, Hayes DN, Perou CM. Comparison of RNA-Seq by poly (A) capture, ribosomal RNA depletion, and DNA microarray for expression profiling. BMC Genomics 2014; 15: 419.
[http://dx.doi.org/10.1186/1471-2164-15-419] [PMID: 24888378]

[82]   Wery M, Descrimes M, Thermes C, Gautheret D, Morillon A. Zinc-mediated RNA fragmentation allows robust transcript reassembly upon whole transcriptome RNA-Seq. Methods 2013; 63(1): 25-31.
[http://dx.doi.org/10.1016/j.ymeth.2013.03.009] [PMID: 23523657]

[83]   Nagalakshmi U, Wang Z, Waern K, *et al.* The transcriptional landscape of the yeast genome defined by RNA sequencing. Science 2008; 320(5881): 1344-9.
[http://dx.doi.org/10.1126/science.1158441] [PMID: 18451266]

[84]   Mortazavi A, Williams BA, McCue K, Schaeffer L, Wold B. Mapping and quantifying mammalian transcriptomes by RNA-Seq. Nat Methods 2008; 5(7): 621-8.
[http://dx.doi.org/10.1038/nmeth.1226] [PMID: 18516045]

[85]   Picelli S, Björklund AK, Reinius B, Sagasser S, Winberg G, Sandberg R. Tn5 transposase and tagmentation procedures for massively scaled sequencing projects. Genome Res 2014; 24(12): 2033-40.
[http://dx.doi.org/10.1101/gr.177881.114] [PMID: 25079858]

[86]   Chen YR, Zheng Y, Liu B, Zhong S, Giovannoni J, Fei Z. A cost-effective method for Illumina small RNA-Seq library preparation using T4 RNA ligase 1 adenylated adapters. Plant Methods 2012; 8(1): 41.
[http://dx.doi.org/10.1186/1746-4811-8-41] [PMID: 22995534]

[87]   Jayaprakash AD, Jabado O, Brown BD, Sachidanandam R. Identification and remediation of biases in the activity of RNA ligases in small-RNA deep sequencing. Nucleic Acids Res 2011; 39(21)e141
[http://dx.doi.org/10.1093/nar/gkr693] [PMID: 21890899]

[88]   Raabe CA, Tang TH, Brosius J, Rozhdestvensky TS. Biases in small RNA deep sequencing data. Nucleic Acids Res 2014; 42(3): 1414-26.
[http://dx.doi.org/10.1093/nar/gkt1021] [PMID: 24198247]

[89]   Quail MA, Otto TD, Gu Y, *et al.* Optimal enzymes for amplifying sequencing libraries. Nat Methods 2011; 9(1): 10-1.
[http://dx.doi.org/10.1038/nmeth.1814] [PMID: 22205512]

[90]   Hoeijmakers WA, Bártfai R, Françoijs KJ, Stunnenberg HG. Linear amplification for deep sequencing. Nat Protoc 2011; 6(7): 1026-36.
[http://dx.doi.org/10.1038/nprot.2011.345] [PMID: 21720315]

[91]   Mohsen MG, Kool ET. The Discovery of Rolling Circle Amplification and Rolling Circle Transcription. Acc Chem Res 2016; 49(11): 2540-50.
[http://dx.doi.org/10.1021/acs.accounts.6b00417] [PMID: 27797171]

[92]   Illumina Seqeuncing https://www.illumina.com/

[93]   IonTorrent Seqeuncing www.thermofisher.com/fr/en/home/brands/ion-torrent.html

[94]   Rothberg JM, Hinz W, Rearick TM, *et al.* An integrated semiconductor device enabling non-optical

genome sequencing. Nature 2011; 475(7356): 348-52.
[http://dx.doi.org/10.1038/nature10242] [PMID: 21776081]

[95]   Laehnemann D, Borkhardt A, McHardy AC. Denoising DNA deep sequencing data-high-throughput sequencing errors and their correction. Brief Bioinform 2016; 17(1): 154-79.
[http://dx.doi.org/10.1093/bib/bbv029] [PMID: 26026159]

[96]   SMRT Seqeuncing https://www.pacb.com/smrt-science/smrt-sequencing/

[97]   Flusberg BA, Webster DR, Lee JH, *et al.* Direct detection of DNA methylation during single-molecule, real-time sequencing. Nat Methods 2010; 7(6): 461-5.
[http://dx.doi.org/10.1038/nmeth.1459] [PMID: 20453866]

[98]   Shendure J, Porreca GJ, Reppas NB, *et al.* Accurate multiplex polony sequencing of an evolved bacterial genome. Science 2005; 309(5741): 1728-32.
[http://dx.doi.org/10.1126/science.1117389] [PMID: 16081699]

[99]   Huang YF, Chen SC, Chiang YS, Chen TH, Chiu KP. Palindromic sequence impedes sequencing-b--ligation mechanism. BMC Syst Biol 2012; 6 (Suppl. 2): S10.
[http://dx.doi.org/10.1186/1752-0509-6-S2-S10] [PMID: 23281822]

[100] Chiu RW, Sun H, Akolekar R, *et al.* Maternal plasma DNA analysis with massively parallel sequencing by ligation for noninvasive prenatal diagnosis of trisomy 21. Clin Chem 2010; 56(3): 459-63.
[http://dx.doi.org/10.1373/clinchem.2009.136507] [PMID: 20026875]

[101] Cloonan N, Forrest AR, Kolle G, *et al.* Stem cell transcriptome profiling *via* massive-scale mRNA sequencing. Nat Methods 2008; 5(7): 613-9.
[http://dx.doi.org/10.1038/nmeth.1223] [PMID: 18516046]

[102] McKernan KJ, Peckham HE, Costa GL, *et al.* Sequence and structural variation in a human genome uncovered by short-read, massively parallel ligation sequencing using two-base encoding. Genome Res 2009; 19(9): 1527-41.
[http://dx.doi.org/10.1101/gr.091868.109] [PMID: 19546169]

[103] Ronaghi M, Uhlen M, Nyren P. A sequencing method based on real-time pyrophosphate. Science 1998; 281: 5.

[104] Oxford Nanopore Technologies https://nanoporetech.com/

[105] Rhee M, Burns MA. Nanopore sequencing technology: nanopore preparations. Trends Biotechnol 2007; 25(4): 174-81.
[http://dx.doi.org/10.1016/j.tibtech.2007.02.008] [PMID: 17320228]

[106] Henley RY, Carson S, Wanunu M. Studies of RNA Sequence and Structure Using Nanopores. Prog Mol Biol Transl Sci 2016; 139: 73-99.
[http://dx.doi.org/10.1016/bs.pmbts.2015.10.020] [PMID: 26970191]

[107] Kulkarni P, Frommolt P. Challenges in the Setup of Large-scale Next-Generation Sequencing Analysis Workflows. Comput Struct Biotechnol J 2017; 15: 471-7.
[http://dx.doi.org/10.1016/j.csbj.2017.10.001] [PMID: 29158876]

[108] Mardis ER. The $1,000 genome, the $100,000 analysis? Genome Med 2010; 2(11): 84.
[http://dx.doi.org/10.1186/gm205] [PMID: 21114804]

[109] Ledergerber C, Dessimoz C. Base-calling for next-generation sequencing platforms. Brief Bioinform 2011; 12(5): 489-97.
[http://dx.doi.org/10.1093/bib/bbq077] [PMID: 21245079]

[110] Ewing B, Green P. Base-calling of automated sequencer traces using phred. II. Error probabilities. Genome Res 1998; 8(3): 186-94.
[http://dx.doi.org/10.1101/gr.8.3.186] [PMID: 9521922]

[111] Ewing B, Hillier L, Wendl MC, Green P. Base-calling of automated sequencer traces using phred. I.

Accuracy assessment. Genome Res 1998; 8(3): 175-85.
[http://dx.doi.org/10.1101/gr.8.3.175] [PMID: 9521921]

[112] Li H, Durbin R. Fast and accurate long-read alignment with Burrows-Wheeler transform. Bioinformatics 2010; 26(5): 589-95.
[http://dx.doi.org/10.1093/bioinformatics/btp698] [PMID: 20080505]

[113] Langmead B, Trapnell C, Pop M, Salzberg SL. Ultrafast and memory-efficient alignment of short DNA sequences to the human genome. Genome Biol 2009; 10(3): R25.
[http://dx.doi.org/10.1186/gb-2009-10-3-r25] [PMID: 19261174]

[114] Myers EW. The fragment assembly string graph. Bioinformatics 2005; 21 (Suppl. 2): ii79-85.
[http://dx.doi.org/10.1093/bioinformatics/bti1114] [PMID: 16204131]

[115] Simpson JT, Durbin R. Efficient de novo assembly of large genomes using compressed data structures. Genome Res 2012; 22(3): 549-56.
[http://dx.doi.org/10.1101/gr.126953.111] [PMID: 22156294]

[116] Aird D, Ross MG, Chen WS, *et al.* Analyzing and minimizing PCR amplification bias in Illumina sequencing libraries. Genome Biol 2011; 12(2): R18.
[http://dx.doi.org/10.1186/gb-2011-12-2-r18] [PMID: 21338519]

[117] Kivioja T, Vähärautio A, Karlsson K, *et al.* Counting absolute numbers of molecules using unique molecular identifiers. Nat Methods 2011; 9(1): 72-4.
[http://dx.doi.org/10.1038/nmeth.1778] [PMID: 22101854]

[118] Islam S, Zeisel A, Joost S, *et al.* Quantitative single-cell RNA-seq with unique molecular identifiers. Nat Methods 2014; 11(2): 163-6.
[http://dx.doi.org/10.1038/nmeth.2772] [PMID: 24363023]

[119] Smith T, Heger A, Sudbery I. UMI-tools: modeling sequencing errors in Unique Molecular Identifiers to improve quantification accuracy. Genome Res 2017; 27(3): 491-9.
[http://dx.doi.org/10.1101/gr.209601.116] [PMID: 28100584]

[120] National Institute of Standards and Technology https://www.nist.gov/

[121] Baker SC, Bauer SR, Beyer RP, *et al.* External RNA Controls Consortium. a progress report. Nat Methods 2005; 2(10): 731-4.
[http://dx.doi.org/10.1038/nmeth1005-731] [PMID: 16179916]

[122] Satija R, Shalek AK. Heterogeneity in immune responses: from populations to single cells. Trends Immunol 2014; 35(5): 219-29.
[http://dx.doi.org/10.1016/j.it.2014.03.004] [PMID: 24746883]

[123] Kolodziejczyk AA, Kim JK, Svensson V, Marioni JC, Teichmann SA. The technology and biology of single-cell RNA sequencing. Mol Cell 2015; 58(4): 610-20.
[http://dx.doi.org/10.1016/j.molcel.2015.04.005] [PMID: 26000846]

[124] Saliba AE, Westermann AJ, Gorski SA, Vogel J. Single-cell RNA-seq: advances and future challenges. Nucleic Acids Res 2014; 42(14): 8845-60.
[http://dx.doi.org/10.1093/nar/gku555] [PMID: 25053837]

[125] Tang F, Barbacioru C, Wang Y, *et al.* mRNA-Seq whole-transcriptome analysis of a single cell. Nat Methods 2009; 6(5): 377-82.
[http://dx.doi.org/10.1038/nmeth.1315] [PMID: 19349980]

[126] Islam S, Kjällquist U, Moliner A, *et al.* Highly multiplexed and strand-specific single-cell RNA 5′ end sequencing. Nat Protoc 2012; 7(5): 813-28.
[http://dx.doi.org/10.1038/nprot.2012.022] [PMID: 22481528]

[127] Wu AR, Wang J, Streets AM, Huang Y. Single-Cell Transcriptional Analysis. Annu Rev Anal Chem (Palo Alto, Calif) 2017; 10(1): 439-62.
[http://dx.doi.org/10.1146/annurev-anchem-061516-045228] [PMID: 28301747]

[128] Haque A, Engel J, Teichmann SA, Lönnberg T. A practical guide to single-cell RNA-sequencing for biomedical research and clinical applications. Genome Med 2017; 9(1): 75.
[http://dx.doi.org/10.1186/s13073-017-0467-4] [PMID: 28821273]

[129] Picelli S, Björklund AK, Faridani OR, Sagasser S, Winberg G, Sandberg R. Smart-seq2 for sensitive full-length transcriptome profiling in single cells. Nat Methods 2013; 10(11): 1096-8.
[http://dx.doi.org/10.1038/nmeth.2639] [PMID: 24056875]

[130] Hashimshony T, Wagner F, Sher N, Yanai I. CEL-Seq: single-cell RNA-Seq by multiplexed linear amplification. Cell Rep 2012; 2(3): 666-73.
[http://dx.doi.org/10.1016/j.celrep.2012.08.003] [PMID: 22939981]

[131] Jaitin DA, Kenigsberg E, Keren-Shaul H, *et al.* Massively parallel single-cell RNA-seq for marker-free decomposition of tissues into cell types. Science 2014; 343(6172): 776-9.
[http://dx.doi.org/10.1126/science.1247651] [PMID: 24531970]

[132] Velten L, Anders S, Pekowska A, *et al.* Single-cell polyadenylation site mapping reveals 3′ isoform choice variability. Mol Syst Biol 2015; 11(6): 812.
[http://dx.doi.org/10.15252/msb.20156198] [PMID: 26040288]

[133] Hashimshony T, Senderovich N, Avital G, *et al.* CEL-Seq2: sensitive highly-multiplexed single-cell RNA-Seq. Genome Biol 2016; 17: 77.
[http://dx.doi.org/10.1186/s13059-016-0938-8] [PMID: 27121950]

[134] Svensson V, Vento-Tormo R, Teichmann SA. Exponential scaling of single-cell RNA-seq in the past decade. Nat Protoc 2018; 13(4): 599-604.
[http://dx.doi.org/10.1038/nprot.2017.149] [PMID: 29494575]

[135] Klein AM, Mazutis L, Akartuna I, *et al.* Droplet barcoding for single-cell transcriptomics applied to embryonic stem cells. Cell 2015; 161(5): 1187-201.
[http://dx.doi.org/10.1016/j.cell.2015.04.044] [PMID: 26000487]

[136] Macosko EZ, Basu A, Satija R, *et al.* Highly Parallel Genome-wide Expression Profiling of Individual Cells Using Nanoliter Droplets. Cell 2015; 161(5): 1202-14.
[http://dx.doi.org/10.1016/j.cell.2015.05.002] [PMID: 26000488]

[137] Chromium System https://www.10xgenomics.com/

[138] Rosenberg AB, Roco CM, Muscat RA, *et al.* Single-cell profiling of the developing mouse brain and spinal cord with split-pool barcoding. Science 2018; 360(6385): 176-82.
[http://dx.doi.org/10.1126/science.aam8999] [PMID: 29545511]

[139] Svensson V, Natarajan KN, Ly LH, *et al.* Power analysis of single-cell RNA-sequencing experiments. Nat Methods 2017; 14(4): 381-7.
[http://dx.doi.org/10.1038/nmeth.4220] [PMID: 28263961]

[140] Ziegenhain C, Vieth B, Parekh S, Reinius B, Guillaumet-Adkins A, Smets M, *et al.* Comparative Analysis of Single-Cell RNA Sequencing Methods. Mol Cell 2017; 65: 631-43.
[http://dx.doi.org/10.1016/j.molcel.2017.01.023]

[141] Harris TD, Buzby PR, Babcock H, *et al.* Single-molecule DNA sequencing of a viral genome. Science 2008; 320(5872): 106-9.
[http://dx.doi.org/10.1126/science.1150427] [PMID: 18388294]

[142] Mamanova L, Andrews RM, James KD, *et al.* FRT-seq: amplification-free, strand-specific transcriptome sequencing. Nat Methods 2010; 7(2): 130-2.
[http://dx.doi.org/10.1038/nmeth.1417] [PMID: 20081834]

[143] Scotti MM, Swanson MS. RNA mis-splicing in disease. Nat Rev Genet 2016; 17(1): 19-32.
[http://dx.doi.org/10.1038/nrg.2015.3] [PMID: 26593421]

[144] Hu Y, Huang Y, Du Y, *et al.* DiffSplice: the genome-wide detection of differential splicing events with RNA-seq. Nucleic Acids Res 2013; 41(2)e39

[http://dx.doi.org/10.1093/nar/gks1026] [PMID: 23155066]

[145] Monlong J, Calvo M, Ferreira PG, Guigó R. Identification of genetic variants associated with alternative splicing using sQTLseekeR. Nat Commun 2014; 5: 4698.
[http://dx.doi.org/10.1038/ncomms5698] [PMID: 25140736]

[146] Goldstein LD, Cao Y, Pau G, *et al.* Prediction and Quantification of Splice Events from RNA-Seq Data. PLoS One 2016; 11(5)e0156132
[http://dx.doi.org/10.1371/journal.pone.0156132] [PMID: 27218464]

[147] Eswaran J, Horvath A, Godbole S, *et al.* RNA sequencing of cancer reveals novel splicing alterations. Sci Rep 2013; 3: 1689.
[http://dx.doi.org/10.1038/srep01689] [PMID: 23604310]

[148] Kremer LS, Bader DM, Mertes C, *et al.* Genetic diagnosis of Mendelian disorders *via* RNA sequencing. Nat Commun 2017; 8: 15824.
[http://dx.doi.org/10.1038/ncomms15824] [PMID: 28604674]

[149] Kang EY, Martin LJ, Mangul S, *et al.* Discovering Single Nucleotide Polymorphisms Regulating Human Gene Expression Using Allele Specific Expression from RNA-seq Data. Genetics 2016; 204(3): 1057-64.
[http://dx.doi.org/10.1534/genetics.115.177246] [PMID: 27765809]

[150] Piskol R, Ramaswami G, Li JB. Reliable identification of genomic variants from RNA-seq data. Am J Hum Genet 2013; 93(4): 641-51.
[http://dx.doi.org/10.1016/j.ajhg.2013.08.008] [PMID: 24075185]

[151] Yan B, Hu Y, Ban KHK, *et al.* Single-cell genomic profiling of acute myeloid leukemia for clinical use: A pilot study. Oncol Lett 2017; 13(3): 1625-30.
[http://dx.doi.org/10.3892/ol.2017.5669] [PMID: 28454300]

[152] Ofengeim D, Giagtzoglou N, Huh D, Zou C, Yuan J. Single-Cell RNA Sequencing: Unraveling the Brain One Cell at a Time. Trends Mol Med 2017; 23(6): 563-76.
[http://dx.doi.org/10.1016/j.molmed.2017.04.006] [PMID: 28501348]

[153] Llorens-Bobadilla E, Zhao S, Baser A, Saiz-Castro G, Zwadlo K, Martin-Villalba A. Single-Cell Transcriptomics Reveals a Population of Dormant Neural Stem Cells that Become Activated upon Brain Injury. Cell Stem Cell 2015; 17(3): 329-40.
[http://dx.doi.org/10.1016/j.stem.2015.07.002] [PMID: 26235341]

[154] Magella B, Adam M, Potter AS, *et al.* Cross-platform single cell analysis of kidney development shows stromal cells express Gdnf. Dev Biol 2018; 434(1): 36-47.
[http://dx.doi.org/10.1016/j.ydbio.2017.11.006] [PMID: 29183737]

[155] Papalexi E, Satija R. Single-cell RNA sequencing to explore immune cell heterogeneity. Nat Rev Immunol 2018; 18(1): 35-45.
[http://dx.doi.org/10.1038/nri.2017.76] [PMID: 28787399]

[156] Shahi P, Kim SC, Haliburton JR, Gartner ZJ, Abate AR. Abseq: Ultrahigh-throughput single cell protein profiling with droplet microfluidic barcoding. Sci Rep 2017; 7: 44447.
[http://dx.doi.org/10.1038/srep44447] [PMID: 28290550]

CHAPTER 9

Immunoelectrophoresis: Recent Advances and Applications

Vinod Singh Gour* and **Ravneet Chug**

Amity Institute of Biotechnology, Amity University Rajasthan, Jaipur, India

Abstract: Antigens and immunoglobulins are important biological entities. Interaction of these components is being considered as a major criterion to assess the health status in veterinary and medical science. Antigen-antibody reactions are highly specific. This interaction between antigen and antibody has been used as the principle of immunoelectrophoresis. This technique has got tremendous applications in basic and applied health science. The present chapter describes principle, process, types, applications, advantages, and limitations of various types of immunoelectrophoresis. This information will be very useful for the students of immunology and scholars who need to develop an understanding of immunoelectrophoresis and its applications.

Keywords: Antigen, Antibody, Crossimmunoelectrophoresis, Disease and diagnosis, Rocket electrophoresis.

INTRODUCTION

The interaction of foreign particle (antigen) in the body of animals triggers the immune system to produce antibodies specific to a particular antigen [1, 2]. This specificity is exploited to detect the presence of a particular antigen in the serum of an individual. The presence and absence of antigen can be detected by a qualitative analysis of serum. The technique used for this purpose is known as Immunoelectrophoresis (IEP). This term was coined by Grabar and Williams, 1953 [3]. Pathogenic proteins can act as an antigen which can stimulate antibodies. In this way, this method becomes very important from disease diagnosis point of view [4]. The details of this technique are being discussed in the following pages.

Principle

The serum proteins (antigens) are separated based on mass and charge using

*** Corresponding author Vinod Singh Gour:** Amity Institute of Biotechnology, Amity University Rajasthan, Jaipur, India; Tel: +91-9414914732; E-mails: vkgaur@jpr.amity.edu; vinodsingh2010@gmail.com

Anupam Jyoti & Neetu Mishra (Eds.)

native protein gel electrophoresis. The separated proteins are then allowed to diffuse against antibodies. The formation of the antigen-antibody complex will lead to precipitation [5]. As the antigen-antibody reaction is a highly specific reaction, so the precipitate formation will indicate the presence of specific antigen in the serum (Fig. **1A**). It is a qualitative detection assay. The formation of antigen-antibody complex takes place when antigen and antibodies are present in optimal proportions, known as antibody titer [5, 6]. If the concentration of either antigen or antibody exceeds this proportion optima, the complex will not be formed. Therefore, the diffusion of antigen from one side and diffusion of antibody from the other side are allowed to facilitate the reaction.

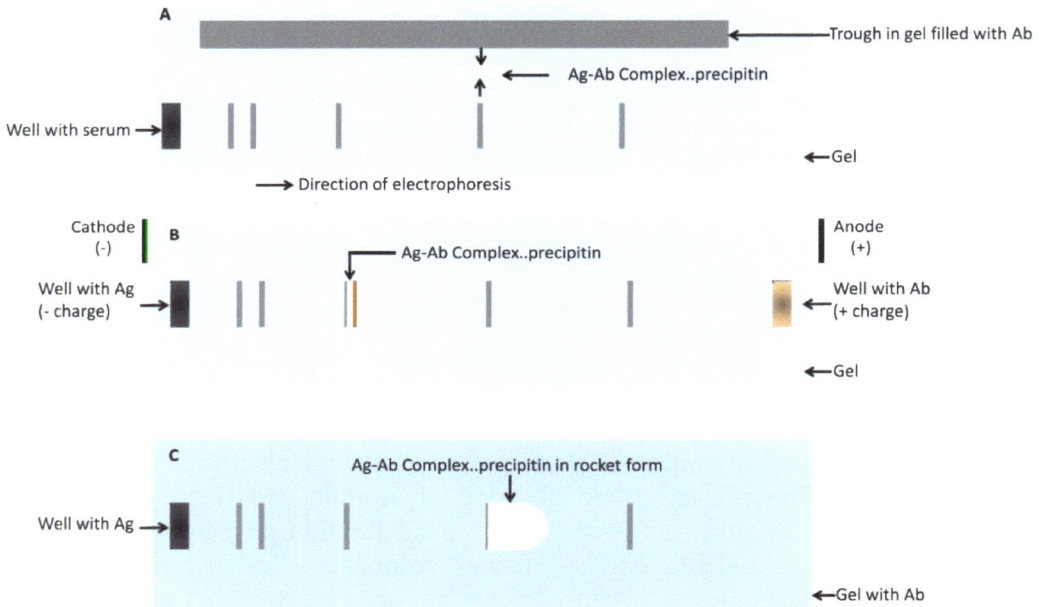

Fig. (1). A. Immunoelectrophoresis, B. Crossimmunoelectrophoresis and C. Rocket immunoelectrophoresis.

Due to the high sensitivity and specific nature of antigen-antibody reaction, this technique is used in the detection of antigens in serum [6]. According to Poretti *et al.* 1999, the sensitivity of IEP was found to range from 58-63% with 97.2% specificity for Ig cystic echinococcosis [7]. It means that it can be used in the diagnosis of any disease and disorders, where antigens-antibodies are involved.

This technique is generally used to detect immunoglobulin G, immunoglobulin M, and immunoglobulin A. However, it has also been used to detect IgD and Ig E [8,

9]. IgD based immunoelectrophoresis has been used in the diagnosis of multiple myeloma [9]. It is more efficient when detecting monoclonal antibodies related antigens, as monoclonal antibodies have a single type of epitope, which interacts with antigen during antigen-antibody complex formation, the presence of IgAλ monoclonal paraprotein has been detected using immunoelectrophoresis in the serum sample with selective IgM deficiency [10].

Based on the requirement of antisera and duration of reaction, IEP has been grouped into two categories: Macro and Micro immunoelectrophoresis [11]. A macro technique requires large quantities of antisera and is time-consuming (14-15 days), however, it is useful to study the interaction of proteins with drugs and hormones. On the other hand, micro technique is less time consuming and requires a very less amount of antisera (100 µl) and antigens (1-3 µl). The procedure of the immunoelectrophoresis is shown in Table **1**.

Applications

a. Pathogen/Protein Detection

IEP has been used to detect various diseases and disorders as it is efficient in detecting antibodies produced by immune system in response to the antigen as presented in Table **2**. According to Morley and Kuslmer, various infections and diseases like rheumatoid arthritis and neoplasia increase the level of C-reactive protein (CRP) in human that can be detected using a very small quantity of antiserum (25-100 µl) and anti-CRP antibodies (0.5 to 1.0 µl) [12, 13].

b. Cancer Diagnosis

This technique also finds its application in the diagnosis of multiple myeloma, a type of bone marrow disease [14]. Akagi and co-workers used on-chip immunoelectrophoresis to detect breast cancer and presented it as a noninvasive "liquid biopsy" for personalized medicine in the future [15].

c. Protease Activity

It has been used to study human IgA protease activity, where the enzyme was produced by more than 100 different bacterial strains of *Streptococcus pneumoniae* and *Haemophilus influenzae* [16].

d. Disorders

It can also be used to detect immunological disorders including autoimmune diseases where the individual produces excess or very low amount of a particular protein.

Table 1. Steps involved in immunoelectrophoresis.

Steps	Activities
1	Take 8 ml of 1.5% agarose in 1 X electrophoresis buffer
2	Spread this suspension on a clean glass slide
3	Mark + and – on both ends of the slide
4	Prepare a ~3 mm well using gel puncture at –ve end of the slide
5	Cut the gel to get a trough as shown in Fig. (**1A**).
6	Add 10-12 µl of Antigen along with Coomassie Brilliant Blue dye solution in the well
7	Place the slide in electrophoresis tub in such a way that –ve end of slide remains close to –ve end of the tub.
8	Set the voltage 50-100V and switch on the current
9	When the dye has travels ¾ length of the slide switch off the current
10	Remove the slide from the tub
11	Add ~200 µl solution of Ab in the trough
12	Incubate it for over-night in a moist chamber in order to diffuse the Ag and Ab
13	Observe the slide carefully
14	Appearance of a milky precipitin arc indicates the presence of particular Ag in given sample.

Table 2. Applications of Immunoelectrophoresis in detection of pathogen/protein

S. No.	Pathogen/Protein	Disease	Reference
1	*Leptospira* spp.	Leptospirosis	[40]
2	*Entamoeba histolytica*	Wide spectrum intestinal diseases	[41]
3	C-reactive protein (CRP)	Rheumatoid arthritis and neoplasia	[12]

e. Differentiation in Protein Profile

Using immunoelectrophoresis, we can differentiate antigenic proteins from non-antigenic ones. For example, 8S myeloma components were found to have no antigenic determinants similar to normal human antigenic determinants [17]. It can also be used to compare proteins of vaginal fluid from normal and hysterectomized women [18]. It means that it can be applied to differentiate two types of samples bound to protein (Ag) profiles. The purity of immunoglobulin isotypes (IgM and IgG) can be ascertained using this technique [19].

Limitations

High concentration (in the order of hundreds of µg/ml) of antibody is required to detect the antigens. The sensitivity of this technique in general ranges between 20 to 200 µg/ml of antibody [20].

Advantages

It is a popular Ag-specific technique commonly used to detect the Ag from an unknown sample.

The process of immunoelectrophoresis has further been strengthened by modifying the procedure leading to the development of techniques like rocket electrophoresis, immunoelectrophoresis 2D and counter current immunoelectrophoresis. These techniques are illustrated in the following pages.

CROSS IMMUNOELECTROPHORESIS

In this version of imunoelectrophoresis, the antigen and antibody are negatively and positively charged respectively. Antigens are loaded in the well near negative electrode and antibodies are loaded in well near positive electrode. The flow of current leads to migration of these Ag and Ab molecules in opposite direction to each other. This process creates better chances of getting the optimal concentration where precipitin is formed (Fig. **1B**). This was first reported by Laurell in 1965 [21]. This technique is known as cross immunoelectrophoresis (CIEP).

Applications

This technique received wide range of applications in disease diagnosis and progress in response to medication and autoimmune disease. The technique is especially useful in forensic science for establishing the origin of body fluids such as blood, semen, and saliva [22]. CIEP can be used to diagnose *Brucella abortus* and *B. melitensis* in infected cattle with 88% and 96% sensitivity respectively which is better than reverse radial immunodiffusion test [23]. Counterimmuno-electrophoresis has also been used to detect *Streptococcus pneumoniae* and *Haemophilus influenzae* in urine samples [24]. Aleutian disease can be diagnosed using CIEP. The infection of parvovirus (Aleutian disease virus -ADV) causes this disease in *Neovison vison* [25]. This test was based on anti-ADV antibodies. The results were compared with those of enzyme-linked immune sorbent assay (ELISA) in terms of specificity and sensitivity. It indicated that both the tests had similar sensitivity, however, specificity of CIEP was found to be slightly higher than ELISA [25].

CIEP has also been applied to test the purity of apolipoproteins and the specificity of antisera isolated from human and rat [26, 27]. After certain modifications, it has been used to detect hydropericardium syndrome virus in poultry birds and the outcomes were compared with reverse passive haem agglutination assay and found to correlate 100% [28].

Even it has an application in the detection of adulteration with pork, beef, and kangaroo meat in heat-processed meat products of chicken [29].

Limitation

During diagnosis of meningitis, it has been reported that there should be a minimum one hundred five colony-forming units per milliliter microbes in cerebrospinal fluid then only it can be detected using CIE [30].

Advantages

CIE has been reported to be a rapid diagnostic tool for detecting *S. pneumoniae* infection in the sputum of the patient suffering from the acute respiratory disease. The time taken was reported to be as less as two hours [31]. It can be used to test the heat-processed meat proteins in order to fix species-specific nature of the meat [29].

ROCKET IMMUNOELECTROPHORESIS (RIE)

The negatively charged antigens are electrophoresed on the gel containing respective antibodies. The Ag-Ab complex will be formed in a rocket (tongue) shape [32]. This technique is used to find out the relative concentration of Ag. The height of the rocket shape is directly proportional to the amount of Ag in the well (Fig. **1C**). If an equal amount of samples derived from two different sources are loaded in two wells separately to compare a particular Ag, then the sample having a higher amount of Ag will show a greater height of Ag-Ab complex (Rocket-shape) in comparison to another one.

Applications

Rocket immunoelectrophoresis has been used in quality assurance/control of vaccines. Ig Y based rocket immunoelectrophoresis has been reported to be a proficient method to monitor pertussis vaccine with about one-fifth of the coefficient of variation [33]. Rocket and 2D immunoelectrophoresis have been used to detect caprine brucellosis induced by *Brucella melitensis* M16 [34]. These methods were found to be superior to traditional tests like culture, Rose Bengal, standard tube agglutination and 2-mercaptoethanol serum agglutination. The RIE test has been used for the qualitative and quantitative assay of specific antigens derived from infectious bursal disease virus (IBDV) in poultry birds [35].

Advantages

Despite the requirement of long run time, rocket immunoelectrophoresis is a method with high sensitivity (78%) and specificity (96%) to detect infectious bursal disease virus-specific antiserum having a low content of antigen [35]. This test was found to be better than the traditional agar gel precipitation test.

Limitation

Three limitations of this technique could be noticed. First, Ag has to be negatively charged. If Ag, under study is neutral or poorly charged then electrophoretic movement of Ag is not possible. Secondly, the sensitivity of this method is about 2 µg/ml of antibody and thirdly, many antigens cannot be dealt at a time by this process [36].

IMMUNOELECTROPHORESIS 2D

First antigens are separated by isoelectrofocusing. Then on second dimension, these separated antigens are run on antibody containing gel, to study the presence of a particular Ag.

Applications

It has been used to understand the production of the C3 and stimulation of complement-dependent neutrophil chemotaxis by S2 (40,000-50,000 D) and S5 (6,000 D) proteins [37]. Nanomaterial mediated response of protein C3 of the complement system was evaluated by modified two dimension immunoelectrophoresis [38].

Advantages

This technique is useful in determining whether a patient produces abnormally low amounts of one or more isotypes of Ig, characteristic of certain immunodeficiency diseases. It can also show whether a patient overproduces some serum proteins, such as albumin, immunoglobulin, or transferrin.

Limitations

Immunoelectrophoresis 2D is a very costly and time-consuming technique.

ADVANCED TECHNIQUES

Classical immunoelectrophoresis has reduced application due to the emergence of more sensitive and specific quantitative immunoelectrophoresis and quantitative radial immunodiffusion [39]. ELISA and radio immunoassay (RIA) have been developed and now in use at a larger scale. Currently, nanomaterial-based protein sensors are being explored in order to develop more specific, highly sensitive, cost-effective and less time-consuming techniques for immune assay.

CONCLUDING REMARKS

Ag-Ab interaction based method of immunoassay (immunoelectrophoresis) has remained a very important technique to detect specific Ag in a given sample. However, techniques derived from immunoelectrophoresis like immunoelectrophoresis 2D, CIEP, and rocket immunoelectrophoresis have got more practical applications and acceptance due to their high sensitivity and specificity.

CONSENT FOR PUBLICATION

Not applicable.

CONFLICT OF INTEREST

The author confirms that this chapter contents have no conflict of interest.

ACKNOWLEDGEMENTS

Authors are thankful to the Co-Editor of this book Dr. Anupam Jyoti, Assistant Professor, Amity Institute of Biotechnology, Amity University Rajasthan, Jaipur for his kind invitation and providing us this opportunity to contribute this chapter.

REFERENCES

[1]   Mariuzza RA, Phillips SE, Poljak RJ. The structural basis of antigen-antibody recognition. Ann Rev Biophys Biophys chem 1987 Jun; 16(1): 139-59.
[http://dx.doi.org/10.1146/annurev.bb.16.060187.001035]

[2]   Casalini S, Dumitru AC, Leonardi F, *et al.* Multiscale sensing of antibody-antigen interactions by organic transistors and single-molecule force spectroscopy. ACS Nano 2015; 9(5): 5051-62.
[http://dx.doi.org/10.1021/acsnano.5b00136] [PMID: 25868724]

[3]   Grabar P, Williams CA. Method permitting the combined study of the electrophoretic and the immunochemical properties of protein mixtures; application to blood serum. Biochim Biophys Acta 1953; 10(1): 193-4.
[http://dx.doi.org/10.1016/0006-3002(53)90233-9] [PMID: 13041735]

[4]   Mehrabani D, Gholami Z, Kohanteb J, Sepehrimanesh M, Hosseini SM. Rocket and Two Dimensional Immunoelectrophoresis in Diagnosis of Caprine Brucellosis. Iran J Public Health 2015; 44(8): 1114-20.

[PMID: 26587475]

[5]   Dean HR, Webb RA. The influence of optimal proportions of antigen and antibody in the serum precipitation reaction. J Pathol Bacteriol 1926; 29(4): 473-92.
[http://dx.doi.org/10.1002/path.1700290412]

[6]   Kanwar SS, Verma ML. Principles and applications of Immuno-diffusion, immuno-electrophoresis, immunofluorescence, ELISA, Western blotting, Minimal Inhibitory Concentration (MIC), Kirby-Bauer method and Widal test 2008.

[7]   Poretti D, Felleisen E, Grimm F, *et al.* Differential immunodiagnosis between cystic hydatid disease and other cross-reactive pathologies. Am J Trop Med Hyg 1999; 60(2): 193-8.
[http://dx.doi.org/10.4269/ajtmh.1999.60.193] [PMID: 10072135]

[8]   Ouaissi A, des Moutis J, Cornette J, Pierce R, Capron A. Detection of IgE antibodies in onchocerciasis using a semi-purified fraction from *Dipetalonema viteae* total antigen. Int Arch Allergy Appl Immunol 1983; 70(3): 231-7.
[http://dx.doi.org/10.1159/000233329] [PMID: 6600714]

[9]   Yavuzer H, Erçalişkan A, Yadigar S, Cengiz M, Döventaş SY, Immunoglobulin D. Multiple Myeloma, A Rare Myeloma Type in Elder Patients: Case Report. Turkiye Klinikleri J Intern Med 2016; 1(2): 103-5.
[http://dx.doi.org/10.5336/intermed.2015-47150]

[10]   Williams SJ, Gupta S. IgA Monoclonal Gammopathy of Undetermined Significant (MGUS) in a Young Woman With Selective IgM Deficiency. J Allergy Clin Immunol 2017; 139(2): AB20.
[http://dx.doi.org/10.1016/j.jaci.2016.12.019]

[11]   MacGillivray AJ. Laboratory techniques in biochemistry and molecular biology. Elsevier 1987.
[http://dx.doi.org/10.1016/0014-5793(87)80455-6]

[12]   Morley JJ, Kushner I. Serum C-reactive protein levels in disease. Ann N Y Acad Sci 1982; 389(1): 406-18.
[http://dx.doi.org/10.1111/j.1749-6632.1982.tb22153.x] [PMID: 6953917]

[13]   Hansson LO, Lindquist L, Linné T, Sego E. Quantitation of C-reactive protein in cerebrospinal fluid and serum by zone immunoelectrophoresis assay (ZIA). J Immunol Methods 1987; 100(1-2): 191-5.
[http://dx.doi.org/10.1016/0022-1759(87)90189-X] [PMID: 3598196]

[14]   Attias P, Moktefi A, Matignon M, *et al.* Monotypic plasma cell interstitial nephritis as the only clinical manifestation in a patient with previously undiagnosed indolent multiple myeloma: A case report. Medicine (Baltimore) 2016; 95(31)e4391
[http://dx.doi.org/10.1097/MD.0000000000004391] [PMID: 27495052]

[15]   Akagi T, Kato K, Kobayashi M, Kosaka N, Ochiya T, Ichiki T. On-chip immunoelectrophoresis of extracellular vesicles released from human breast cancer cells. PLoS One 2015; 10(4)e0123603
[http://dx.doi.org/10.1371/journal.pone.0123603] [PMID: 25928805]

[16]   Mulks MH, Kornfeld SJ, Plaut AG. Specific proteolysis of human IgA by *Streptococcus pneumoniae* and *Haemophilus influenzae*. J Infect Dis 1980; 141(4): 450-6.
[http://dx.doi.org/10.1093/infdis/141.4.450] [PMID: 6989925]

[17]   Johansson SG, Bennich H. Immunological studies of an atypical (myeloma) immunoglobulin. Immunology 1967; 13(4): 381-94.
[PMID: 4168094]

[18]   Raffi RO, Moghissi KS, Sacco AG. Proteins of human vaginal fluid. Fertil Steril 1977; 28(12): 1345-8.
[http://dx.doi.org/10.1016/S0015-0282(16)42982-1] [PMID: 590545]

[19]   Jaton JC, Ungar-Waron H, Sela M. Comparison of IgG and IgM antibodies specific for uridine. Eur J Biochem 1967; 2(1): 106-14.
[http://dx.doi.org/10.1111/j.1432-1033.1967.tb00114.x] [PMID: 4169689]

[20]   Kindt TJ, Goldsby RA, Osborne BA, Kuby J. Kuby immunology. 6th ed., India Freeman book 2008.

[21]   Laurell C-B. Antigen-antibody crossed electrophoresis. Anal Biochem 1965; 10: 358-61.
[http://dx.doi.org/10.1016/0003-2697(65)90278-2] [PMID: 14302464]

[22]   Estela LA, Heinrichs TF. Evaluation of the counterimmunoelectrophoretic (CIE) procedure in a clinical laboratory setting. Am J Clin Pathol 1978; 70(2): 239-43.
[http://dx.doi.org/10.1093/ajcp/70.2.239] [PMID: 29483]

[23]   Ducrotoy MJ, Muñoz PM, Conde-Álvarez R, Blasco JM, Moriyón I. A systematic review of current immunological tests for the diagnosis of cattle brucellosis. Prev Vet Med 2018; 151: 57-72.
[http://dx.doi.org/10.1016/j.prevetmed.2018.01.005] [PMID: 29496108]

[24]   Azmi F, Iqbal J, Nomani N, Ishaque Z, Iqbal J, Rab A. Counter immuno electrophoresis for the diagnosis of *Strept. pneumonia* and *H. influenzae* pneumonia. J Pak Med Assoc 1987; 37(6): 148-52.
[PMID: 3114510]

[25]   Dam-Tuxen R, Dahl J, Jensen TH, Dam-Tuxen T, Struve T, Bruun L. Diagnosing Aleutian mink disease infection by a new fully automated ELISA or by counter current immunoelectrophoresis: a comparison of sensitivity and specificity. J Virol Methods 2014; 199: 53-60.
[http://dx.doi.org/10.1016/j.jviromet.2014.01.011] [PMID: 24462658]

[26]   Holmquist L. Two-dimensional immunoelectrophoresis of human serum very low density apolipoproteins. FEBS Lett 1980; 111(1): 162-6.
[http://dx.doi.org/10.1016/0014-5793(80)80783-6] [PMID: 7358155]

[27]   Dolphin PJ. Two-dimensional immunoelectrophoresis of rat serum apolipoproteins. Electrophoresis 1981; 2(2): 113-6.
[http://dx.doi.org/10.1002/elps.1150020208]

[28]   Manzoor S, Ur-Rahman S, Khan IA. Identification and Titration of Hydro Pericardium Syndrome Virus (HPSV) by using Modified Counter-Current Immuno-Electrophoresis (MCCIE). J Antivir Antiretrovir 2017; 9: 052-4.

[29]   Necidova L, Rencova E, Svoboda I. Counter immunoelectrophoresis: a simple method for the detection of species-specific muscle proteins in heat-processed products. Vet Med (Praha) 2002; 47(5): 143-7.
[http://dx.doi.org/10.17221/5818-VETMED]

[30]   Rytel MW. Counterimmunoelectrophoresis: a diagnostic adjunct in clinical microbiology 1980; 2;11(10): 655-8.
[http://dx.doi.org/10.1093/labmed/11.10.655]

[31]   Spencer RC, Savage MA. Use of Counter and rocket immunoelectrophoresis in acute respiratory infections due to Streptococcus pneumoniae. J Clin Pathol 1976; 29(3): 187-90.
[http://dx.doi.org/10.1136/jcp.29.3.187] [PMID: 5466]

[32]   Ressler N. Electrophoresis of serum protein antigens in an antibody-containing buffer. Clin Chim Acta 1960; 5: 359-65.
[http://dx.doi.org/10.1016/0009-8981(60)90140-6] [PMID: 13856286]

[33]   Matheis W, Schade R. Development of an IgY-based rocket-immunoelectrophoresis for identity monitoring of Pertussis vaccines. J Immunol Methods 2011; 369(1-2): 125-32.
[http://dx.doi.org/10.1016/j.jim.2011.04.013] [PMID: 21586289]

[34]   Mehrabani D, Gholami Z, Kohanteb J, Sepehrimanesh M, Hosseini SM. Rocket and Two Dimensional Immunoelectrophoresis in Diagnosis of Caprine Brucellosis. Iran J Public Health 2015; 44(8): 1114-20.
[PMID: 26587475]

[35]   Raj GD, Thangavelu A, Elankumaran S, Koteeswaran A. Rocket immunoelectrophoresis in the diagnosis of infectious bursal disease. Trop Anim Health Prod 2000; 32(3): 173-8.

[http://dx.doi.org/10.1023/A:1005287732214] [PMID: 10907288]

[36]   Kindt TJ, Goldsby RA, Osborne BA, Kuby J. Kuby immunology Sixth edition India Freeman book. 2008.

[37]   Anjolette FA, Leite FP, Bordon KC, *et al.* Biological characterization of compounds from *Rhinella schneideri* poison that act on the complement system. J Venom Anim Toxins Incl Trop Dis 2015; 21(1): 25.
       [http://dx.doi.org/10.1186/s40409-015-0024-9] [PMID: 26273286]

[38]   Coty JB, Varenne F, Vachon JJ, Vauthier C. Serial multiple crossed immunoelectrophoresis at a microscale: A stamp-sized 2D immunoanalysis of protein C3 activation caused by nanoparticles. Electrophoresis 2016; 37(17-18): 2401-9.
       [http://dx.doi.org/10.1002/elps.201500572] [PMID: 27387591]

[39]   Nerenberg ST, Peetoom F. Use of immunoelectrophoresis and immunodiffusion in clinical medicine. CRC Crit Rev Clin Lab Sci 1970; 1(2): 303-50.
       [http://dx.doi.org/10.3109/10408367009021492] [PMID: 5006216]

[40]   Thresiamma KC, Biju A, Chaurasia R, *et al.* Proteinuria in early detection of human leptospirosis. IJRMS 2017; 5(2): 646-52.

[41]   Savanat T, Chaicumpa W. Immunoelectrophoresis test for amoebiasis. Bull World Health Organ 1969; 40(3): 343-53.
       [PMID: 5306620]

SUBJECT INDEX

www.ingramcontent.com/pod-product-compliance
Lightning Source LLC
Chambersburg PA
CBHW041712210326
41598CB000C7B/627